Introduction to Stereochemistry

Chemistry Student Guides

Editor-in-Chief:
Julie Macpherson, *University of Warwick, UK*

Series editors:
Dudley Shallcross, *University of Bristol, UK*
Paul Taylor, *University of Leeds, UK*
Simon Humphrey, *University of Texas at Austin, USA*

Addressing a challenging concept in undergraduate chemistry, each book in the *Chemistry Student Guides* series aims for the reader to achieve a 'light bulb' moment. Once a concept is understood, the reader will no longer need to use rote memorisation techniques, but instead will have the confidence to see how the learning can be applied across a range of problems, in order to obtain the correct answers.

Each book received student input so it targets exactly where students struggle. Essential student involvement, combined with high editorial engagement from the Board, means that each title is closely tuned to audience needs. The books can be used as a supplement or alternative to core textbooks, and will also work well as a revision tool during exam time.

Titles in the series:

1: Introduction to Stereochemistry
2: Introduction to Contextual Maths in Chemistry

How to obtain future titles on publication:
A standing order plan is available for this series. A standing order will bring delivery of each new volume immediately on publication.

For further information please contact:
Book Sales Department, Royal Society of Chemistry, Thomas Graham House, Science Park, Milton Road, Cambridge, CB4 0WF, UK
Telephone: +44 (0)1223 420066, Fax: +44 (0)1223 420247,
Email: booksales@rsc.org
Visit our website at www.rsc.org/books

Introduction to Stereochemistry

By

Andrew Clark
University of Warwick, UK
Email: A.J.Clark@warwick.ac.uk

Russ Kitson
University of Warwick, UK
Email: R.Kitson@warwick.ac.uk

Nimesh Mistry
University of Leeds, UK
Email: N.Mistry@leeds.ac.uk

and

Paul Taylor
University of Leeds, UK
Email: P.C.Taylor@leeds.ac.uk

with student co-authors

Matthew Taylor
University of Warwick, UK

Michael Lloyd
Imperial College London, UK

and

Caroline Akamune
University of Warwick, UK

ROYAL SOCIETY
OF **CHEMISTRY**

Chemistry Student Guides, No. 1

Print ISBN: 978-1-78801-315-4
EPUB ISBN: 978-1-83916-182-7
Print ISSN: 2632-9867
Electronic ISSN: 2632-9875

A catalogue record for this book is available from the British Library.

The Royal Society of Chemistry is a charity, registered in England and Wales, Number 207890, and a company incorporated in England by Royal Charter (Registered No. RC000524), registered office: Burlington House, Piccadilly, London W1J 0BA, UK, Telephone: +44 (0) 20 7437 8656.

Visit our website at www.rsc.org/books

Printed in the United Kingdom by CPI Group (UK) Ltd, Croydon, CR0 4YY, UK

"The precise arrangement of atoms in 3-dimensional space – stereochemistry – is a fundamental, overarching concept in chemistry. It impinges on our very existence because all the molecules of life (proteins, nucleic acids, carbohydrates, lipids) possess stereochemical features.

However, students often struggle to understand stereochemistry, and find the terminology confusing. This book addresses such issues head on, and teaches the core concepts of stereochemistry in a logical and extremely clear manner. It familiarises students with the language and structural basis of stereochemistry, and, importantly, gives them confidence to draw accurate representations.

The involvement of student co-authors is an added bonus since it ensures that the explanation of difficult concepts is clear, and adequately addresses topics that they find more challenging."

Chris Moody
Sir Jesse Boot Professor, University of Nottingham, UK

Author Biographies

Andrew Clark is a Professor in Chemistry and the University Academic Director responsible for all undergraduate programmes at the University of Warwick.

Russ Kitson is an Associate Professor in Chemistry at the University of Warwick. Russ' research centres around organic chemistry and chemical education, with a focus on inclusive practice, laboratory learning, active learning, authentic learning, game-based learning and employability.

Nimesh Mistry is a Senior Teaching Fellow at the University of Leeds. Nimesh's research interests are in chemistry education, with a focus on laboratory education, organic chemistry education and authentic research experiences.

Paul Taylor is currently Professor of Chemical Education at the University of Leeds, where he is also Pro-Dean for Student Education in the Faculty of Engineering and Physical Sciences. Paul is a National Teaching Fellow.

Matthew Taylor is a third-year undergraduate student of chemistry with medicinal chemistry at the University of Warwick. His research interests lie in synthetic organic chemistry and finding intuitive and creative ways of communicating the subject.

Mike Lloyd is a PhD student at Imperial College London, specialising in single-molecule chemistry for the study of neurodegenerative disease. He graduated from the University of Leeds with an MChem in medicinal chemistry in 2018, which included a summer internship with Dr Nimesh Mistry.

Caroline Akamune is a third-year undergraduate student of chemistry at the University of Warwick. Following a summer research project with Dr Manuela Tosin, she has developed a keen interest in synthetic organic chemistry and chemical biology.

Preface

Welcome to Chemistry Student Guides: *Introduction to Stereochemistry*! This 'student guide' is the product of a partnership between a group of undergraduate students and experienced academic staff from the Universities of Leeds and Warwick, both institutions that have a proud tradition of staff and undergraduate students working in collaboration to co-create knowledge.

This book is absolutely in the partnership tradition. Our team worked together to identify the need for a new text on stereochemistry, decided what to include, and leave out, designed the presentation of the topic and spent many hours discussing the format and style of the chapters.

At the heart of our approach is the tutorial style of teaching that is so important to us in the Department of Chemistry at Warwick and the School of Chemistry at Leeds. Students and their teachers sit down together to grapple with challenging problems in chemistry, break them down into their component parts and work through them collaboratively to gain new understanding.

The result is a book that is truly co-designed and co-authored by our student–staff team. We're grateful for the encouragement we have had from other students and teachers at our Universities and our colleagues at the Royal Society of Chemistry. We hope you enjoy the book and truly welcome feedback on our work and the Chemistry Student Guides concept.

Andrew Clark
Russ Kitson
Nimesh Mistry
Paul Taylor
Matthew Taylor
Michael Lloyd
and
Caroline Akamune

How to Use This Book

Throughout the book you will find useful tips, succinct summaries of a vital points and plenty of exercises to check your understanding. Look out for the key learning points and tips in the margins and please do make sure you 'pass' tests at each checkpoint before progressing. Use the icons to navigate between topics when you need to look back for something.

- ▙ Checkpoint[†]

- 💬 Tip[‡]

- ❗ Key learning point[§]

- ❓ Question[¶]

[†]Icon flag by Viktor Vorobyev from the Noun Project.
[‡]Icon comment by Alice Design from the Noun Project.
[§]Icon exclamation mark by Louis Buck from the Noun Project.
[¶]Icon question mark by José Campos from the Noun Project.

Glossary

Terms included in the glossary are indicated in bold at their first occurrence in the main book and also if they occur in the glossary within other entries.

Term	Definition
Absolute stereochemistry	The precise spatial arrangement of **substituents** around a **stereogenic centre**.
Achiral	An object that is **superimposable** on its mirror image is said to be achiral.
anti	On the opposite side to an atom or group on another carbon atom when the chain is drawn **zig-zag**.
Anomeric effect	A preference for heteroatomic **substituents** adjacent to a heteroatom that is part of a ring to adopt the **axial**, rather than the **equatorial**, position.
Anticlinal	Of a **conformation**, **eclipsed** but not **periplanar**.
Atomic number	The number of protons in the nucleus of a given element.
Atropisomer	A **conformer** that can be isolated as a separate chemical species and arises from restricted rotation about a single bond.
Axial	Used to describe groups bonded to cyclic structures where the bonds are perpendicular to the plane of the ring.
Bidentate	Of a ligand, attached to the metal atom of a coordination complex by two interactions.
Boat conformation	A **conformation** of cyclohexanes and their derivatives with a boat shape.
Chair conformation	An important **conformation** of cyclohexanes and their derivatives with a chair shape.
Chiral	An object (*e.g.* a molecule) that is non-**superimposable** on its mirror image is said to be chiral.
CIP rules	Cahn–Ingold–Prelog rules, which can be used in assignment of configurational descriptors.
cis/trans stereoisomerism	Stereoisomerism due to differing configurational arrangements of two groups that can be on the same or opposite sides of a molecule.
Configuration	The spatial arrangement of atoms in a molecule that can only be altered by breaking and making bonds. Configurational **isomers** can be *e.g.* **enantiomers** or **diastereoisomers**.
Conformation	The spatial arrangement of atoms in a molecule that can be altered by rotation about one or more single bonds and does *not* require breaking and making bonds. **Conformational isomers** are known as **conformers**.

Term	Definition
Conformational isomers	Species that are interchangeable *via* rotation around one or more single bonds (**conformations**) and that lie at an energy minimum.
Conformers	Species that are interchangeable *via* rotation around one or more single bonds (**conformations**) and that lie at an energy minimum.
Coupling constant	A measure, usually in hertz, of the interaction between two NMR-active nuclei.
Decalin	A bicyclic compound comprising two fused cyclohexane rings.
Diastereoisomer	Diastereoisomers are stereoisomers that are non-**superimposable**, non-mirror images.
Diastereotopic	A relationship between two groups in a molecule where replacement of one of the two groups *vs.* the other gives rise to a pair of **diastereoisomers**.
1,3-Diaxial interactions	Of atoms or groups in **axial** positions of a **chair conformation**, the interactions such groups have with similar groups four bonds away (*e.g.* on carbon atoms 1 and 3).
Dihedral angle	The angle between two planes. Also called the torsion angle when referring to two substituents separated by three bonds.
Eclipsed	Having **substituents** at the closest spatial distance from **substituents** on an adjacent centre, so the **dihedral angle** is zero.
Enantiomer	A pair of enantiomers is a pair of stereoisomers that are non-**superimposable** mirror images of one another.
Enantiotopic	A relationship between two groups in a molecule where replacement of one group *vs.* the other group gives rise to a pair of **enantiomers**.
Equatorial	Used to describe groups bonded to cyclic structures where the bonds lie approximately in the plane of the ring.
E/Z stereochemistry	Stereoisomerism due to differing configurational arrangements of groups in alkenes, defined using the Cahn–Ingold–Prelog (**CIP**) rules.
Facial (*fac*)	Of an **octahedral** MA_3B_3 coordination complex, when the three A ligands (and B ligands) lie on a face of the octahedron.
Fiducial point	A reference point added to a diagram to aid visualisation.
Flip method	A way of manipulating structural drawings to aid assignment of configuration, in which an assignment is flipped according to a protocol.
Functionality (functional group)	Groups of atoms within a molecule that have characteristic properties.
Furanose	A carbohydrate containing a five-membered ring with four carbon atoms and one oxygen atom.
Gauche	Of a **conformation**, **staggered** but not **periplanar**. See also **synclinal**.
Geminal	Of two groups, on the same carbon atom.
Geometric isomers	**Isomers** that differ in the spatial arrangement of atoms or groups with respect to a bond where rotation is restricted.
Half-chair	A **conformation** of cyclohexane and derivatives with part of a **chair** shape and five carbon atoms in the same plane.

Term	Definition
Hashed	By convention, a bond drawn hashed is projecting away from the viewer.
Homotopic	A relationship between two groups in a molecule where replacement of one group *vs.* the other gives the same compound.
Hybridisation	A combination of orbitals on the same atom.
Inversion	Of a pyramidal sp^3 hybridised centre with a lone pair, a switch in configuration passing through a planar sp^2 hybridised transition state.
Isomers	Molecules that have the same molecular formula but have different line or stereochemical formulae and so have different properties.
Karplus relationship	A mathematical model of how **coupling constants** vary with **dihedral angle**.
Magnetic equivalence	Where nuclei are indistinguishable by NMR spectroscopy.
Meridional (*mer*)	Of an **octahedral** MA_3B_3 coordination complex, when the three A ligands (and B ligands) lie in a plane with each other and the central metal atom.
meso compound	A compound that contains two or more stereogenic atoms but that is **superimposable** on its mirror image and therefore **achiral**.
Monodentate	Of a ligand, attached to the metal atom of a coordination complex by one interaction.
Newman projection	A way of drawing a molecule to represent the relative positions of bonds on two adjacent atoms. The viewer imagines looking along the bond between the two atoms in question.
Octahedral	Of a coordination complex with six ligands, when all the ligands are approximately equidistant from each other and as far apart as possible.
Optical activity	The way in which non-racemic **chiral** molecules rotate a plane of polarised light.
Optical rotation	The rotation of a plane of polarised light by a non-racemic, **chiral** sample, usually measured in a polarimeter.
Orthogonal	Perpendicular.
Periplanar	Of a conformation, in the same plane as. May be *syn* or *anti*.
Pi bond (π bond)	Bond formed through lateral overlap of lobes of p orbitals on adjacent atoms.
Pseudoaxial	As **axial**, but used when discussing rings or cyclic transition states that are not in true **chair conformations**.
Pseudochirality	Apparent chirality that arises from two otherwise identical **substituents** having opposite **configurations**.
Pseudoequatorial	As **equatorial**, but used when discussing rings or cyclic transition states that are not in true **chair conformations**.
Pyranose	A carbohydrate containing a six-membered ring with five carbon atoms and one oxygen atom.
Racemic mixture (racemate)	A racemic mixture or racemate is a 1:1 mixture of **enantiomers**.
Relative stereochemistry	The spatial arrangement of a **substituent** (*e.g.* at one **stereogenic centre**) relative to a **substituent** at another atom *e.g.* on the same side (**syn**) or the opposite side (**anti**).
Resonance	A way of representing a structure as the sum of localised contributory structures.

Term	Definition
Sawhorse projection	A way of drawing a molecule to represent the relative positions of bonds on two adjacent atoms. The bond between the two key atoms is drawn as a diagonal from lower left to upper right. One of the bonds to each of the two key atoms is then drawn as vertical, these two pointing towards each other. Remaining bonds are drawn at 120°.
Sigma bond (σ bond)	Bond formed through overlap of orbitals lying along the axis of the bond.
Square planar	Of a four-coordinate complex, when the metal and four ligands all lie in the same plane with the ligands approximately equidistant.
Square pyramidal	Of a five-coordinate complex, when four of the ligands all lie in the same plane with the ligands equidistant and a fifth ligand lies out of the plane to form a pyramid with a square base.
Staggered	Having **substituents** at the maximum spatial distance from **substituents** on an adjacent centre.
Stereogenic axis (axis of chirality)	An axis about which a set of groups is held so that it results in a spatial arrangement that is not **superimposable** on its mirror image.
Stereogenic centre (stereocentre)	For atoms with tetrahedral geometry, an atom that has four different **substituents** attached to it.
Stereochemistry	A topic in chemistry concerned with the spatial arrangement of atoms in a molecule and how the spatial arrangement affects structure and reactivity.
Stereoisomers	**Isomers** that have the same connectivity between atoms, including the same bond multiplicities, and so differ only in the spatial arrangement of their atoms.
Substituent	An atom or group *other than a hydrogen atom* that is bonded to an atom of interest.
Superimposable	Of two objects (*e.g.* molecules), identical, so that an image of one object maps exactly onto an image of the other.
Swap method	A way of manipulating structural drawings to aid assignment of **configuration**, in which **substituents** are swapped according to a protocol.
syn	On the same side as an atom or group on another carbon atom when the chain is drawn **zig-zag.**
Synclinal	Of a **conformation**, **staggered** but not **periplanar**. See also **gauche**.
Theoretical modelling	A way to use a theory to make predictions about a system.
Torsional strain	The increase in energy resulting from the unfavourable interaction of electrons in two separate bonds.
Tridentate	Of a ligand, attached to the metal atom of a coordination complex by three interactions.
Trigonal bipyramidal	Of a coordination complex, when the metal and three of the five ligands all lie in the same plane with the ligands equidistant and the fourth and fifth ligand lie out of the plane on opposite sides to form two fused pyramids with a common triangular base.
Twist-boat	A **conformation** of cyclohexane and derivatives that resembles a twisted **boat conformation**.
VSEPR	Valence Shell Electron Pair Repulsion theory.

Term	Definition
Vicinal angle	Used in NMR spectroscopy with the same meaning as **dihedral angle**.
Wedged	By convention, a bond drawn as a bold wedge is projecting towards the viewer.
Zig-zag	By convention, the main chain of a molecule is drawn in the plane of the page so that the chain extends continually across the page, creating a zig-zag pattern.

Table of Contents

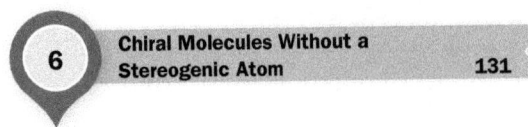

By the End of This Chapter You Will:

☐ Start to recognise the importance of stereochemistry.
☐ Start to understand what stereoisomerism means.
☐ Gain a feeling for the ways we draw and describe stereoisomers.
☐ Understand how to use this book.

What You Will Get from This Chapter

This short chapter is really just to set the scene for the rest of the book, highlighting the importance of stereochemistry and the approaches we can use to study the topic. Don't worry if the new material appears daunting at this point. We'll go through it slowly in the relevant chapters. Perhaps the most important bit of Chapter 1 is the final section, Section 1.4, explaining 'how to use this book'.

Stereochemistry

1.1 Thalidomide – Why Stereochemistry Is Important

In the 1960s, thalidomide was an over-the-counter drug that was given to pregnant women, mostly in Germany, to treat morning sickness. The use of this drug led to 10 000 babies being born with phocomelia (deformed limbs). Approximately half of the children survived and have had to live with this condition.

Representing the structure of thalidomide as shown in Figure 1.1 does not explain why this happened, because it gives the impression that it is a flat molecule. Importantly, one of the carbon atoms (marked *) is bound to four different groups. This means that two types of **isomer** can form, based on the 3D orientation of the four groups.

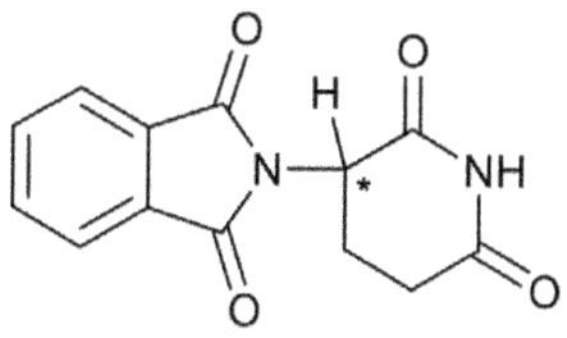

Figure 1.1 Planar representation of thalidomide.

A better representation to show this 3D effect is to represent the bonds as **wedged** to show groups pointing towards you, and as **hashed** to show groups pointing away, as in Figure 1.2. Here we

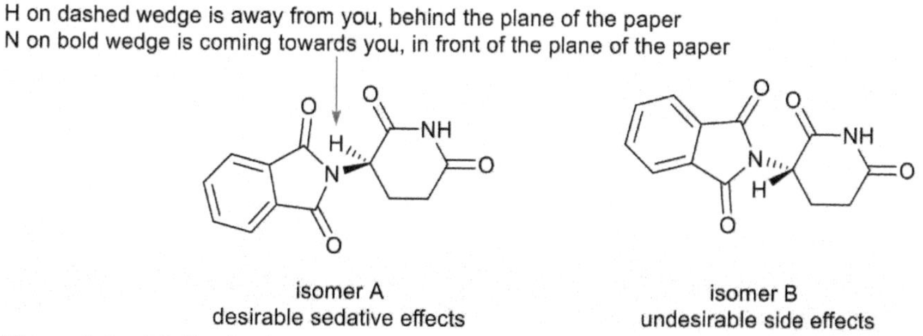

H on dashed wedge is away from you, behind the plane of the paper
N on bold wedge is coming towards you, in front of the plane of the paper

isomer A
desirable sedative effects

isomer B
undesirable side effects

Figure 1.2 2D 'hashed and wedged' representation of thalidomide shows that it exists as two isomers.

can see the two isomers that thalidomide can have. It turns out that the isomer on the right is the one responsible for phocomelia. If the company that produced the drug had been aware of this then it is possible that they could have sold the drug as only the bioactive isomer (although interconversion of the isomers in the body, known as racemisation, might still have complicated matters).

This story, perhaps more than any other in history, signifies why it is important to have an understanding of molecules in 3D. This topic is known as **stereochemistry** and throughout this book we will introduce the main concepts in this area.

1.2 Stereoisomerism

You may well have come across types of isomerism, such as structural isomerism, where the atoms in two molecules with the same molecular formula are joined together in different ways. This book will introduce another type of isomerism, called **stereoisomerism**. This is when the atoms are connected in the same way but they adopt different positions in space.

Once you've understood what kinds of stereoisomers there are, it's important to consider how readily they can interconvert. If they basically can't, we have a set of configurational isomers that we may even be able to separate and study separately. Even if they can interconvert, we may be able to observe different **conformational isomers** by spectroscopy – understanding which isomer is of lowest energy can be helpful in predicting certain properties of molecules, including how they react.

1.2.1 Configurational Isomerism

Configurational isomers (Figure 1.3) are stereoisomers that cannot interconvert readily. We will explain how these types of isomers can be seen in alkenes, tetrahedral carbons, planar organic molecules and inorganic complexes.

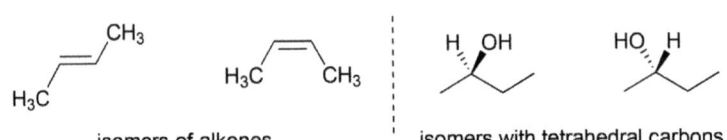

isomers of alkenes   isomers with tetrahedral carbons

Figure 1.3   Examples of configurational isomers.

Configuration as a topic is covered in Chapter 2 (alkenes), Chapter 3 (tetrahedral carbons), Chapter 6 (helicenes, biphenyls and allenes) and Chapter 7 (inorganic complexes).

1.2.2 Conformational Isomerism

Any organic molecule that can have rotation about a bond can exist as conformational isomers (Figure 1.4). This topic relates to the different spatial arrangements in 3D that molecules can have through the rotation of bonds.

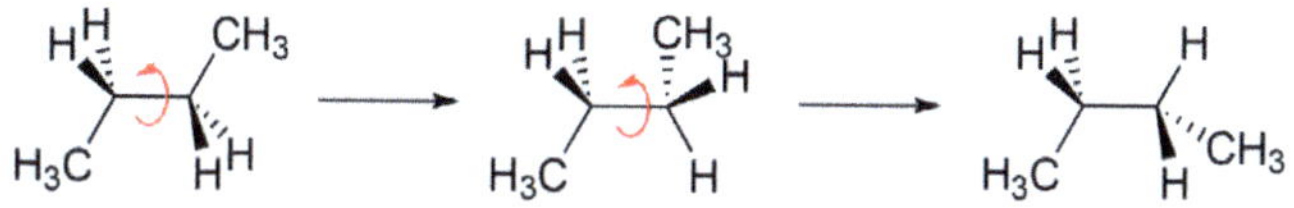

Figure 1.4   Example of conformations.

We'll be studying the **conformations** of acyclic (no ring structures) and cyclic compounds in Chapters 4 and 5, respectively.

Of course, all this requires us to represent the stereochemical information in the chemical structures we draw, as in Figures 1.2–1.4. This isn't easy, but we'll guide you through it step by step.

> Don't worry if Figure 1.4 isn't making much sense at the moment. It's difficult to visualise features like this, which is why we've written this book!

1.3 Representing Molecules in 3D

As we've said, as we discuss different concepts of stereochemistry and, essentially, the 3D structure of molecules, we will be introducing the different ways of drawing and naming molecules that represent their 3D character. One of the most unfamiliar and tricky areas is representing conformations, but we'll get there!

1.3.1 New Ways of Drawing Molecules

A **Newman projection** is the perspective obtained by looking down a specific bond in a molecule. It is used most often with alkanes where the central C–C bond is considered. The C atom closer to you has the lines for the bonds running to it and the further C atom is represented by a circle, as shown in Figure 1.5. Newman projections are useful in explaining effects such as the **Karplus**

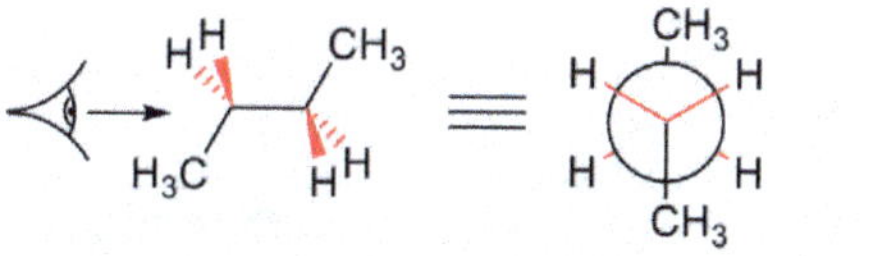

Newman projection

Figure 1.5   Newman projection.

relationship in NMR spectroscopy, and to help us identify the lowest-energy conformations of molecules that arise from steric and electronic interactions.

Chair conformations (Figure 1.6) represent a more accurate view of what a six-membered ring, such as a cyclohexane, looks like in 3D. By looking at chair conformations, we can tell what the most stable form of a particular six-membered ring is. This can then allow us to predict what will form from a reaction, and hence how we can control the stereochemistry of the products.

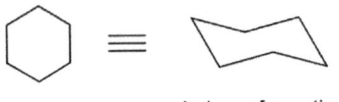

chair conformation

Figure 1.6 Chair conformation.

1.3.2 Stereochemical Terminology

There will be some new jargon used when talking about stereo-chemistry, and a lot of the words are similar in meaning, so it can get quite confusing. Table 1.1 gives a brief summary of stereo-chemical nomenclature and why we use these terms. We've also included a glossary of the difficult terms we've used in the book. Throughout the book, if you see text in **bold**, it means you will find a definition in the glossary.

Table 1.1 Terminology used in stereochemistry.

Nomenclature	Purpose	Explanation
R/S	Configuration of a single **stereocentre**	Assigned to 'right-' and 'left-hand' **enantiomers**. These come from the Latin for right (*rectus*) and left (*sinister*).
+/−	**Optical activity** of the whole molecule	In what direction a molecule rotates plane-polarised light.
syn/anti	**Relative stereo-chemistry** of two groups	Used to refer to groups being on the same (*syn*) or opposite (*anti*) sides with free rotation.
cis/trans	Relative stereo-chemistry between two **substituents**	Used to indicate whether substituents are on the same side (*cis*) or the opposite (*trans*) side with restricted rotation.
E/Z	Alkenes	Indicates whether the highest-priority groups on each end of a double bond are on the same (*Z*) or opposite sides (*E*).

1.4 How to Use This Book

Imagine you are in a small-group teaching session. Your tutor or instructor invites you to grapple with the topic yourself, to make molecular models, to try various exercises, to take control of your learning, with their support. This book is designed to model that way of teaching and learning, together. So please follow the suggestions, make models, try out the concepts and so on.

In the same session, your tutor or instructor would break each topic down into component concepts, emphasising key learning points, giving top tips and checking that you had 'got' that component before moving on to the next stage. This book reproduces this way of learning. Look out for the key learning points (❶) and tips (💬) in the margins and please do make sure you 'pass' tests at each checkpoint (⚑) before progressing. Each chapter starts with a checklist of key concepts that you need to really 'get'. These are highlighted at the checkpoints through the chapter and recapped in the chapter conclusion.

Stereochemistry is a challenging but rewarding topic. You can't just read about it and hope to understand. Please engage and enjoy. Good luck!

💬

Use the key learning point boxes to navigate between chapters when you need to look back for something.

By the End of This Chapter You Will:

- [ ] Understand the bonding in an alkene.
- [ ] Understand why some alkenes exhibit stereochemistry and that different geometric isomers have different properties.
- [ ] Be comfortable with when and how to assign *E/Z* and *cis/trans* prefixes to describe an alkene.
- [ ] Know the Cahn–Ingold–Prelog (CIP) rules to assign the stereochemistry of alkenes.
- [ ] Be comfortable with using NMR spectroscopy to elucidate the structure of an alkene.

What You Will Get from This Chapter

Visualising molecules in 3D is hard. Fortunately, chemists use quite a few of the important stereochemical concepts and terms to describe the shapes of more 2D molecules, like alkenes. So, even if you really came to this book to learn about chirality and conformations in molecules with sp^3 carbon atoms, we suggest you venture first into the sp^2 world of alkenes. You can build your confidence in describing groups as *cis* and *trans* to each other, following the Cahn–Ingold–Prelog (CIP) rules and using NMR spectroscopy to observe stereochemical features. In short, Chapter 2 is a great place to start your exploration of stereochemistry.

Alkenes

Alkenes have applications throughout our day-to-day lives. Consider the tomato (Figure 2.1). After being picked, tomatoes are often ripened off the vine synthetically using an alkene called ethylene (ethene). The ripening process makes tomatoes their distinctive red colour; that pigment is yet again an alkene: lycopene. Even now, while reading about alkenes you are using an alkene, retinal, which allows perception of light. Alkenes are special because of the rigidity and arrangement of **functional groups** around the alkene carbon atoms. We refer to this as stereochemistry and it is a major part of a chemist's understanding of the world around us.

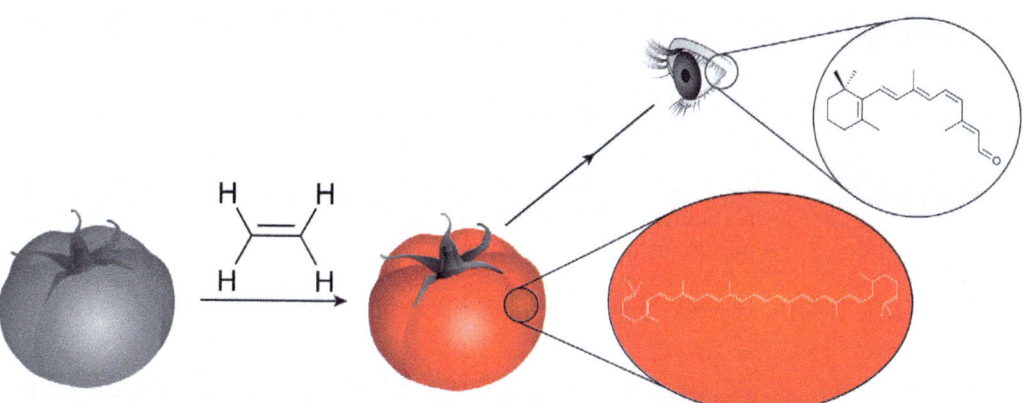

Figure 2.1   Alkenes are all around us, even requiring an alkene in the eye to perceive them!

2.1 Structure of an Alkene Bond

One of the simplest chemical **functionalities** is the alkene. Its structure is drawn as in Figure 2.2. This representation is intended to show the double bond nature of the alkene. The double bond is significantly stronger than a carbon–carbon single bond, which is the only C–C bond found in alkanes. This strength is a result of the orbital interactions between the two alkene carbon atoms, which we will discuss in the next section. The substituents, R, are oriented at around 120° to each other. This is understood using **theoretical models** (specifically hybridisation theory and valence shell electron pair repulsion theory, abbreviated as **VSEPR**[†]).

[†]You will come across these models in the first or second year of an undergraduate course.

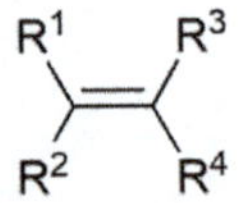

$$R^1 \diagup\!\!=\!\!\diagdown R^3$$
$$R^2 \diagdown \quad \diagup R^4$$

Figure 2.2   General alkene skeletal depiction with four generic substituents, R.

In the structure shown in Figure 2.3, the carbon atom has three hybridised sp^2 orbitals and one unhybridised p orbital. This is the type of carbon that makes up an alkene.

lone p orbital

↓

C ⟩ 3 sp^2 orbitals

Figure 2.3   An isolated sp^2 hybridised carbon with three sp^2 orbitals oriented at 120° to one another and a singular p orbital **orthogonal** to them.

Hybridisation is a method of mixing the atomic orbitals of a carbon atom in such a way that we produce a set of orbitals that are often more practical to work with. In this text we will mostly consider sp^2 and sp^3 carbons (pronounced ess-pee-two and ess-pee-three, respectively). We won't go into the derivation of hybridisation but will focus on the result of it.

> ❗ **Key Learning Point**
>
> Hybridisation is a method of mixing the atomic orbitals of the carbon in such a way that we produce a set of orbitals that are often more practical to work with.

For this chapter we will focus on the sp^2 carbon (Figure 2.4) but in further chapters the sp^3 carbon (Figure 2.5) will be more prominent.

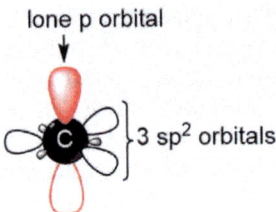

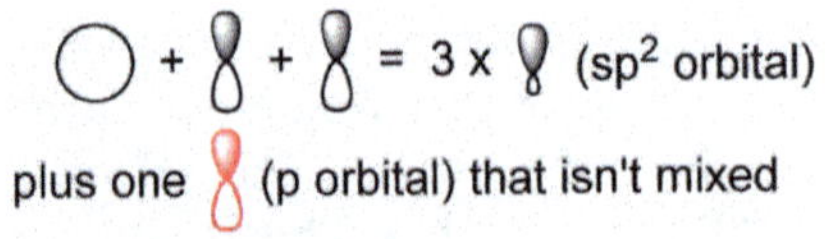

plus one ⬗ (p orbital) that isn't mixed

Figure 2.4   How an s orbital and only two of the three p orbitals are mixed to produce sp^2 hybridisation.

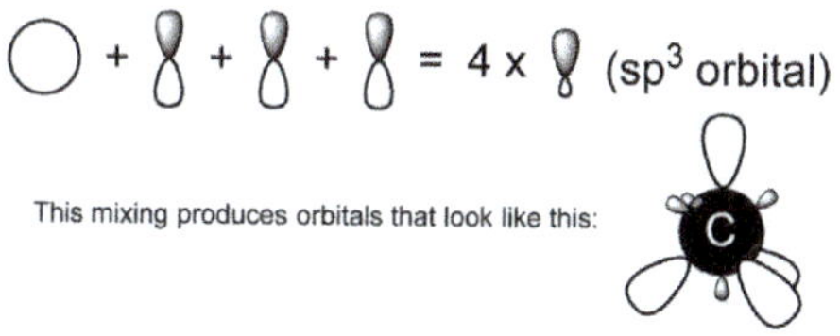

$$\bigcirc + \; \text{\textonequarter} + \; \text{\textonequarter} + \; \text{\textonequarter} = 4 \times \; \text{\textonequarter} \; (\text{sp}^3 \text{ orbital})$$

This mixing produces orbitals that look like this:

Figure 2.5   How the sp³ orbitals are mixed; notice that this time we mixed the lone p orbital, making all the hybrid orbitals sp³ orbitals.

2.1.1 Alkenes Are Made of π Bonds

After looking at Figure 2.2, one could ask, 'Why do we draw two bonds when there is only one net interaction between the two carbon atoms?' The answer is based in the fact that there are two interactions between the sp² carbons; these interactions result from their hybridisation.

The two drawn bonds in Figure 2.2 display the presence of a **σ bond** and a **π bond**. These bond types are based on their geometries. A σ bond consists of a single electron cloud along the interatomic or longitudinal axis, arising as a result of the interaction of the two interatomic sp² orbitals. The π bond, in contrast, sits above and below this axis as this is where the two p orbitals interact, as shown in Figure 2.6.

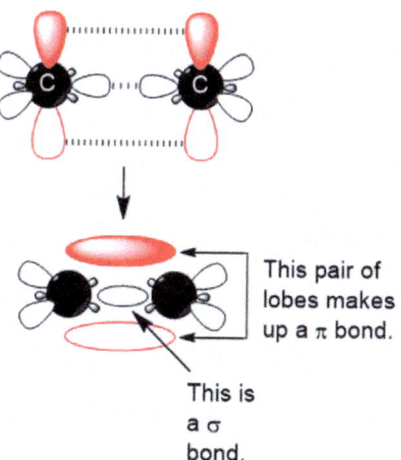

This pair of
lobes makes
up a π bond.

This is
a σ
bond.

Figure 2.6   How the longitudinal sp² and p orbitals interact to form the σ and π bonds, respectively.

2.1.2 *Alkene double bonds Cannot Rotate*

Now, let's consider twisting our alkene double bond. Hold the left carbon atom in place and slowly turn the right one 180°. What would happen with regards to the bonding? Consider how the orbital overlap would be affected.

As we start to get towards orthogonality (after twisting by 90°, Figure 2.7) we reduce the π orbital overlap to zero. This means that the p orbitals are not parallel and so can't interact in a way that forms a bond. We have reduced the bond character to one rather than two by breaking one of the bonds in the alkene. This all means that the substituents on the alkene can't twist or rotate around the C–C bond without the input of a significant energetic force (260 kJ mol^{-1})!

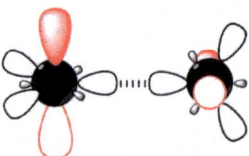

Figure 2.7   After twisting by 90°, we find that the p orbitals are orthogonal and have zero net bonding character as the p orbitals don't overlap in a constructive or destructive way. This only leaves the σ bond between the carbons.

Restricted bond rotation is the reason why we see stereochemistry in alkenes, but the principle can be applied to some other restricted bonds. A common example is in an amide where the N–C bond is restricted by an effect called **resonance**. Generally, bond rotation is described as restricted if it requires more than 100 kJ mol^{-1} to overcome. We will come back to this later, in Section 4.6.

2.2 Geometric Isomerism in Alkenes

So, as shown in Figure 2.8, alkenes have geometric isomerism. **Geometric isomers** only differ in the spatial arrangement of groups on either side of a restricted bond. This results in the arrangement of substituents with respect to the double bond being significant for the properties of the molecule.

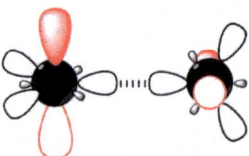

Figure 2.8   Substituents can't change position on an alkene without significant energetic requirements.

2.2.1 *Isomers Are Non-superimposable*

Let's consider a general alkene with four different substituents, R^1 to R^4. What happens if we imagine swapping R^3 and R^4. Is it the same molecule? In other words, are the isomers superimposable? Take a look at Figure 2.9.

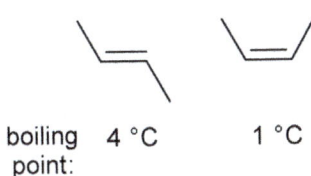

Figure 2.9 The two alkene isomers are sat on top of each other. Notice how you can't lay them perfectly on top of one another as R^3 and R^4 don't match up. The molecules are described as non-superimposable.

That's right, we actually find that they aren't superimposable. They are unique isomers.

2.2.2 *Isomers Have Different Properties*

By now it is clear that alkenes can exist as unique geometric isomers, but the way in which their properties differ based on their configuration may not be obvious. The simplest example of this is but-2-ene. The boiling point of the molecule is affected by the configuration (Figure 2.10). This can be rationalised by imagining that the molecules on the left pack together better. In reality a lot of different effects come into play, but for now this analogy will suffice.

> **■ Checkpoint**
>
> You should now understand why some alkenes exhibit stereochemistry and that different geometric isomers have different properties.

boiling point: 4 °C   1 °C

Figure 2.10 The boiling points of the two geometric isomers of but-2-ene are different.

2.3 Alkene Stereochemistry Nomenclature

In the world of chemistry, we need to know what we are looking at and how to communicate that information. Consider the molecule in Figure 2.11. How could we communicate its structure?

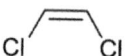

Figure 2.11   One of the two geometric isomers of 1,2-dichloroethene.

There are two systems in place for an alkene stereochemistry description: the *cis/trans* system and the *E/Z* system. Many chemists prefer the *E/Z* system as it is more thorough but the *cis/trans* system is a little more intuitive. Try to get comfortable with both as they are essential descriptors for any chemist.

2.3.1   Cis *and* trans *Stereochemistry*

> ❗ **Key Learning Point**
>
> The *cis/trans* naming system is only used when each carbon atom in an alkene is bonded to one hydrogen each!

In Figure 2.11, we can see that, relative to the double bond, the two chlorine substituents in 1,2-dichloroethene are on the same side. Chemists have a system for naming alkenes (and other restricted geometric isomers) where the positions of two groups (not considering hydrogen) are compared. This system is called *cis/trans* stereochemistry. We denote the relative stereochemistry with the prefixes '*cis*' and '*trans*', which mean that the substituents are on the 'same side' and 'opposite side' of the double bond, respectively. One way of remembering this is by considering that a *trans*atlantic ship sails to the 'other side' of the Atlantic. One should note that this system is *only used when each of the alkene carbons is bonded to a hydrogen.*

We established that our 1,2-dichloroethene had its substituents on the same side of the double bond. This then means that we can describe our molecule as *cis*-1,2-dichloroethene. Now, let's try it in reverse. What would *trans*-1,2-dichloroethene look like?

We know that our basis is ethene. Now it has two chlorine atoms in the 1 and 2 positions, and they are on opposite sides of the double bond. Therefore, our *trans*-1,2-dichloroethene corresponds to Figure 2.12.

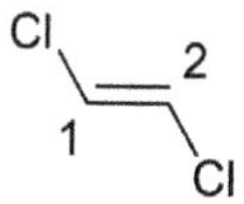

Figure 2.12   *Trans*-1,2-dichloroethene.

To check your understanding, have a go at deciding what *cis/trans* prefix you would assign to the molecules in Figures 2.13 and 2.14.

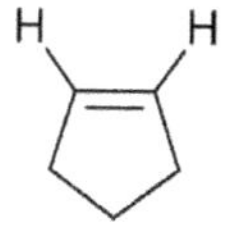

Figure 2.13   Molecule 2.3.1a for assignment.

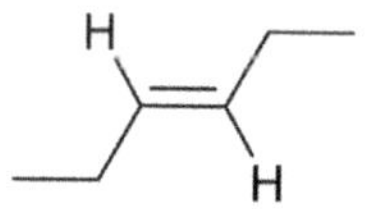

Figure 2.14   Molecule 2.3.1b for assignment.

Although we have not looked at a cyclic alkene before, the same principles apply! We can see that the two 'substituents' (non-hydrogen atoms) bonded to our alkene carbons are on the same side, thus we assign molecule 2.3.1a as '*cis*'. For molecule 2.3.1b, we can see that the non-hydrogen substituents are attached to two carbons on opposite sides and so we would assign it as '*trans*'.

2.3.2   E and Z Stereochemistry

In the last subsection, we encountered relative stereochemical nomenclature, noting that we should only use it when we have a hydrogen atom on each of the alkene carbons. Let's consider a molecule where this is not the case, as shown in Figure 2.15.

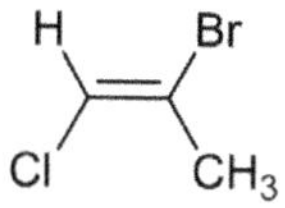

Figure 2.15   1-Chloro-2-bromopropene.

Since we don't have a hydrogen on each carbon, we can no longer use the *cis/trans* stereochemistry nomenclature! We need a new system that investigates the local environment of each carbon. This system is called *E/Z* stereochemistry and utilises Cahn–Ingold–Prelog (**CIP**) rules. This system assigns the configuration of each alkene carbon atom based on the priority of the groups around it.

Firstly, let's define the CIP rules. After learning the CIP rules, we will then apply them to assigning alkenes using the *E/Z* system. In later chapters, we will use them to assign other kinds of stereochemistry.

1. Priority is based on **atomic number**; higher atomic numbers gain higher priorities. With respect to alkenes, priority groups are what we compare when looking across the double bond, similar to how we assigned groups in the *cis/trans* system. For example, oxygen (atomic number = 8) has a higher priority than hydrogen (atomic number = 1).

   Priority is written like positions in a race: 1 is highest, 2 is lower, 3 is lower than that, and so on.

2. If we have two groups with equal priority bonded to our alkene carbon atoms, we perform a tiebreak. To do this we look around each of these atoms (usually carbons), looking for the highest-priority atom bonded to it. The highest-priority bonded atom is compared between the two carbons and the carbon that wins the tiebreak will get higher priority.

   Rule 2 is reapplied if the tie continues; this may be the case for alkyl chains bonded to the alkene.

   Think about the molecule shown in Figure 2.16. You can see that the atoms directly bonded to the alkene are both carbons. This means we have to perform a tiebreak. We look around both carbons and compare their substituents; as they are identical methylene carbons, the tiebreak continues for each carbon in the chains. That is, until the last carbon, where we see the carbon of the CH_2Cl group (circled in red) is bonded to a higher-priority atom (Cl, atomic number = 17) relative to the O (atomic number = 8) of the

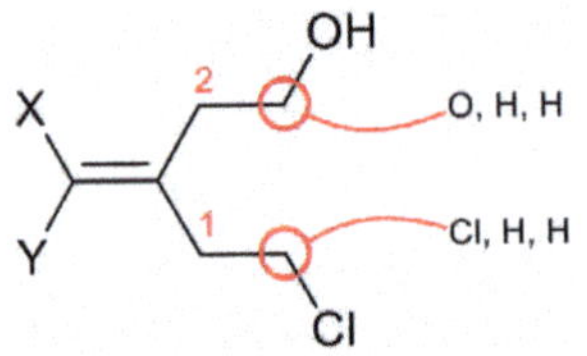

Figure 2.16   The bottom group takes priority as the tiebreak was won by the Cl. The competing carbons have been circled and the label indicates the atoms they are bonded to. As you can see, the Cl takes priority over the O.

CH$_2$OH group (also circled in red). Remember that you are only comparing the highest atoms, not groups, so it is O, not OH. This means the bottom chain ultimately won the tiebreak so takes priority, as seen in Figure 2.16. In the second worked example, you will see a way to organise this on paper.

❗ Key Learning Point

Remember that it's just the atomic number of the bonded atom that we consider, so a large phenyl group has a lower priority than a chlorine or even a fluorine atom!

---

3. If an atom has multiple bonds to another (*e.g.* C=C, C=O or C≡C) then those bonds count as such. So a C≡C would constitute three bonds to a carbon, as you might expect. You will see more examples of this later in this section. In Figure 2.17, the circled carbons are being compared. They are both bonded to an oxygen, but as the double bond constitutes two bonds to oxygen it takes higher priority than the single bond to oxygen of the alcohol. The atoms bonded to the respective carbon are shown in order of priority (*i.e.* O, O, C). We thus label the top carbon as '1' and the lower carbon as '2', in accordance with Rule 1.

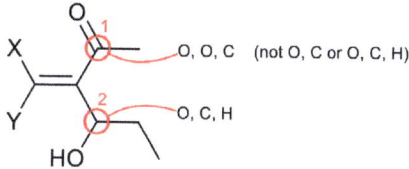

Figure 2.17 Comparison of ketones to secondary alcohols, the C=O bond constitutes two bonds to oxygen, not one! This result means that the top carbon will take higher priority. Be sure to watch out for this, as many students can trip up on this in an exam!

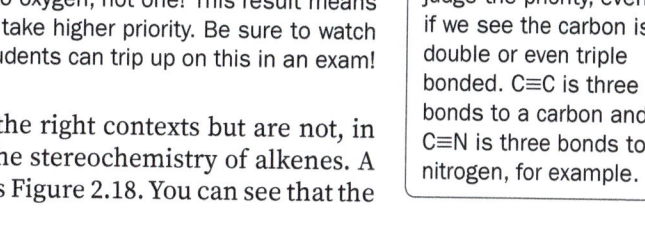

Remember that we should have three atoms for each carbon that we are using to judge the priority, even if we see the carbon is double or even triple bonded. C≡C is three bonds to a carbon and C≡N is three bonds to nitrogen, for example.

Cis and *trans* are useful terms in the right contexts but are not, in general, a good way to describe the stereochemistry of alkenes. A good example to make this point is Figure 2.18. You can see that the

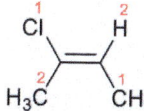

Figure 2.18 Although the methyl groups are *cis* to each other, the molecule is *not* a *cis*-alkene. We will see that the alkene stereochemistry is best described using *E/Z* nomenclature.

two methyl groups are on the same side of the double bond, like a *cis* label. However, if we try assigning priorities, we will find that the higher-priority substituents (Cl and CH_3) are now on opposite sides!

---

❶ Key Learning Point

Remember that it's okay to say that identical groups are *cis* or *trans* to each other. However, this is very different to calling the whole molecule *cis* or *trans*, as you can see with Figure 2.18.

Just remember only to use the *cis/trans* system to label an alkene if both carbons have a hydrogen atom bonded to them, *i.e.* Figure 2.19. For any other case, we use *E* or *Z* to describe the stereochemistry using the Cahn–Ingold–Prelog (CIP) rules.

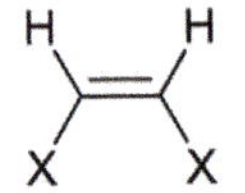

Figure 2.19   General structure of an alkene that can be named using the *cis/trans* system (*cis* in this case). The X groups can be any group other than H.

◼ Checkpoint

You should now understand how to assign *E/Z* and *cis/trans* prefixes to describe an alkene and when it is useful to use *cis/trans*.

These concepts will be reiterated many times throughout this text, so don't worry if you make any mistakes! Now we know the CIP priority rules, we can now start applying it to *E/Z* assignments! To get to grips with this new system we will run through some examples, starting with Figure 2.20.

Let's look at the circled carbon first (Figure 2.20).

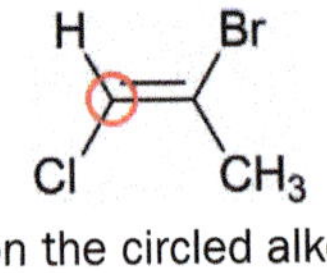

Figure 2.20   We will focus on the circled alkene carbon first.

Consider the priorities of the substituents. The atomic number of H is 1 and the atomic number of Cl is 17, therefore Cl takes priority 1 and H takes priority 2 (Figure 2.21).

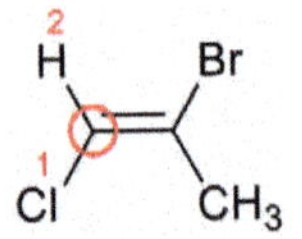

Figure 2.21   We have worked out the priority of the substituents around our first alkene carbon.

We don't need to assign a priority number to the other alkene carbon, just to the two alkene substituents. Now, let's look at this second alkene carbon, circled in Figure 2.22.

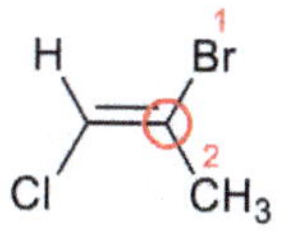

Figure 2.22   Focusing on the circled carbon, we find that the priority groups are as indicated.

The atomic number of Br is 35 and the atomic number of C is 6, therefore Br takes priority 1 and C takes priority 2. Remember that we only consider atoms directly bonded to our carbon, for now we ignore the hydrogens on the CH_3 group. Let's take a step back and look at the whole molecule after we have assigned all the substituents (Figure 2.23).

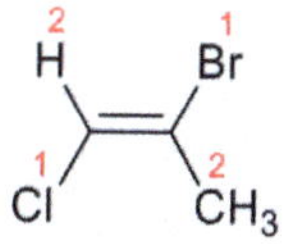

Figure 2.23   An overview of our molecule after CIP priority assignment.

For alkenes, we now look at how the higher-priority groups are arranged about the double bond. In our case, they are on opposite sides. How do we communicate this in the E/Z stereochemical system?

In the E/Z stereochemical system we use the prefixes 'Z' and 'E', meaning 'on the same side' and 'on the opposite side' respectively. This is the same as for the *cis/trans* system. However, the E/Z system means that we are referring to the arrangement of the higher-priority groups, clearing up any ambiguity in the naming of a configuration.

Going back to our example (Figure 2.23), we see that the priority groups are on opposite sides; therefore, we will assign our molecule as (*E*)-1-chloro-2-bromopropene. Now, let's work backward. How would we draw (*Z*)-1-chloro-2-bromopropene? We just need to have the priority groups on the same side, so we need only

The names are rooted in German, '*Z*' meaning *zusammen* or 'together' and '*E*' meaning *entgegen* or 'opposite' (Figure 2.24). Some students remember the distinction with the mnemonic 'Enemies are on opposite sides.'

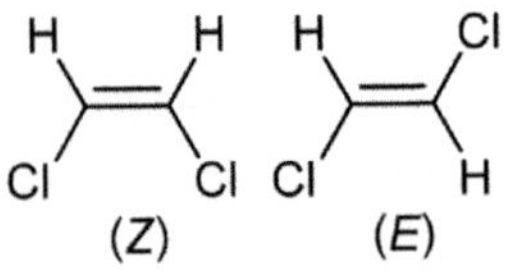

Figure 2.24 (*E*)-1,2-dichloroethene and (*Z*)-1,2-dichloroethene. This figure shows how *E*/*Z* maps onto the *cis*/*trans* system. Remember that the *cis*/*trans* system could only work here because both carbons have a hydrogen bonded to them!

swap the substituents on the second alkene carbon, producing the desired molecule (Figure 2.25)!

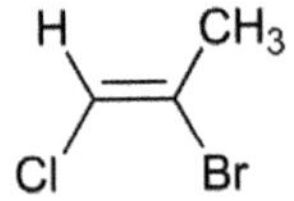

Figure 2.25 (*Z*)-1-chloro-2-bromopropene.

Our second example is shown in Figure 2.26. We will ignore the full IUPAC‡ name as it is quite complicated.

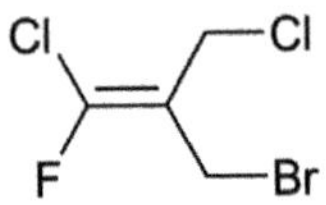

Figure 2.26 Example stereochemical assignment.

The left-hand alkene carbon (Figure 2.27) is the best place to start. First, let's find what the priorities of the groups attached to carbon are.

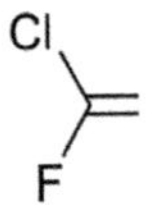

Figure 2.27 Left half of the molecule.

The Cl (atomic number = 17) takes top priority and will as such be labelled '1' and the F (atomic number = 9) will thus be labelled '2' (Figure 2.28).

‡IUPAC is the organisation that has standardised naming practices in chemistry. IUPAC names are precise descriptors, but for the purpose of these exercises they are not needed. We only really care for the prefixes in this context.

Figure 2.28   Left half of the molecule after priority assignment.

On the right-hand side (Figure 2.29), we have a trickier problem, the two atoms bonded to the carbon are tied; they are both carbons.

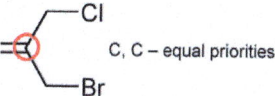

Figure 2.29   Right half of the molecule. Our alkene is bonded to two carbons; this means we have a tiebreak to carry out.

We therefore go along one carbon on each chain to break the tie. What we do is list the bonded groups for both carbons in order of priority, as seen in Figure 2.30. When comparing the priorities, we see that Br (atomic number = 35) takes a higher priority than Cl (atomic number = 17).

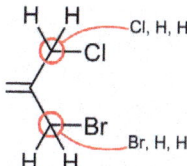

Figure 2.30   The atoms bonded to the tied carbons are indicated. We can see that the bottom branch will take priority.

Overall, this means that we label the priorities as shown in Figure 2.31.

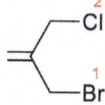

Figure 2.31   Priority assignment of the right half of the molecule.

We can now fully assign the configuration of our molecule (Figure 2.32)! As the priority substituents are on the opposite side, we assign the alkene as *E*! As in 'enemies are on opposite sides'.

Do not just eyeball the methylenebromide and methylenechloride substituents and say that the methylenebromide takes priority! Many students get into this habit and will struggle to solve more complex stereochemical assignments as they don't follow the CIP rules properly. Take care and be thorough!

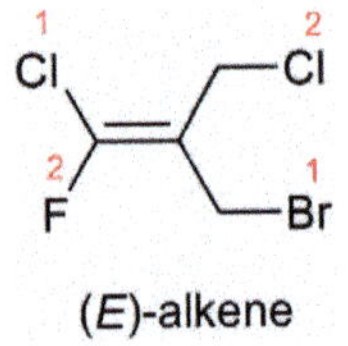

Figure 2.32 Full priority assignment with configuration labelled.

Our next example, Figure 2.33, will reinforce the concepts we have covered so far. It certainly does look intimidating but remember we only look at one half of the molecule at a time. By doing this you can observe, in a concise manner, many of the places that people trip up.

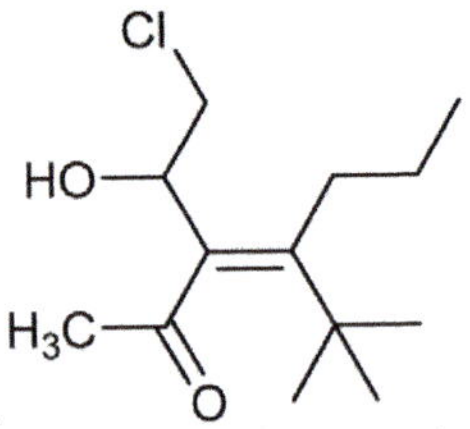

Figure 2.33 Further example for stereochemical assignment.

Let's start with the right-hand side; this side should be relatively simple to assign. You are encouraged to give both halves a go before you read the explanations. At this point we will try to make the examples a bit tricky so you can learn from any potential mistakes and see where you could be tripped up in an exam. Let's dive right into the bonded atom labelling!

Again, we have a tiebreak to resolve! Looking at the tied carbons we can label the bonded atoms and compare, as seen in Figure 2.34.

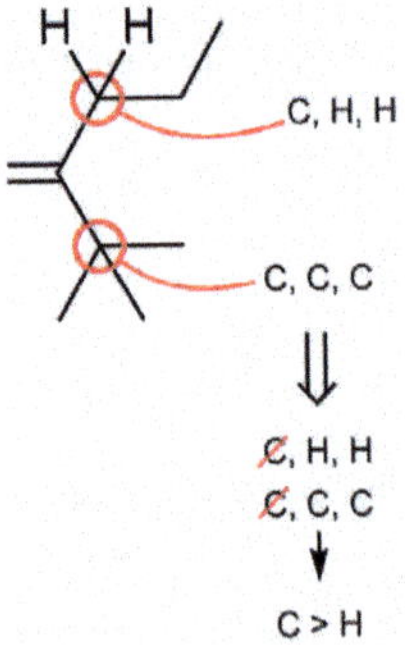

Figure 2.34 The atoms bonded to the tied carbons are indicated. We then compare the highest-priority bonded atom; here they are the same. This means we consider the second highest, where C is compared against H. C gives the bottom branch the priority.

We can see that the *t*-butyl branch takes priority over the *n*-propyl branch. The highest-priority bonded atom for both is a carbon so we look at the second highest priority, a comparison between C and H, where C wins the tiebreak.

As a result, we can assign the right-hand side, as shown in Figure 2.35.

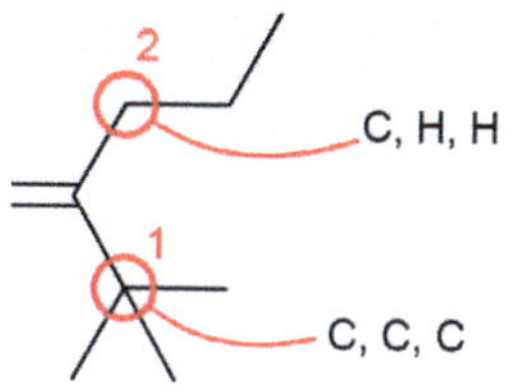

Figure 2.35 The right-hand side of the molecule after stereochemical assignment.

Next, let's look at the left side (Figure 2.36)! The molecular substituents are chosen as ones that you can easily trip up on, so have a go. It should be a fun challenge!

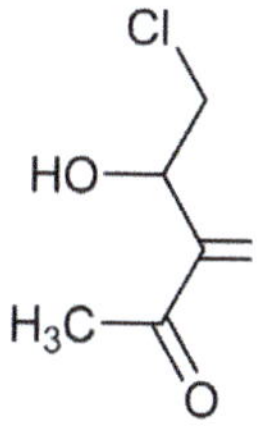

Figure 2.36 The left half of the molecule.

As always, let's label the bonded atoms to the tied carbons. The top tied carbon is bonded to O, C and H (Figure 2.37), but what about the lower one?

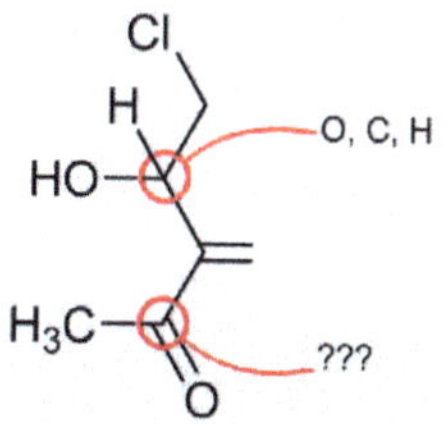

Figure 2.37 The top circled carbon has its bonded atom priorities labelled. But what about the bottom circled carbon?

If we go back to the CIP rules, we see that a ketone counts as two bonds to an O, so we label the carbon as O, O, C (Figure 2.38).

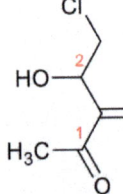

Figure 2.38   A C=O constitutes two carbon–oxygen bonds. Again, we can compare the bonded atoms and find that the bottom branch takes priority.

> One of the best tips an undergraduate can receive is to always consider the hydrogen atoms on a molecule. The fact they are implicit makes it easy to forget they are there! Draw them in. No one minds if you draw them on a molecule in an exam; who knows, you may get credit for spotting them!

As before, we see that the highest-priority atom of each carbon is the same; this time, O. So, to discern the priorities we look at the second-highest priority atoms: O and C. Here O wins priority for the bottom branch. A common mistake is to say that the C=O is one bond, resulting in the top branch being given priority (thanks to the Cl)! The left side is consequently assigned (Figure 2.39) and then we can work out the configuration.

Figure 2.39   Left half of the molecule after stereochemical assignment.

As the priority groups are on the same side that means we can assign the configuration as *Z* (Figure 2.40)!

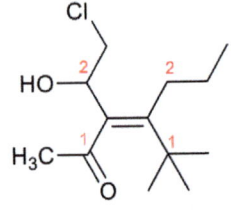

(*Z*)-alkene

Figure 2.40   (*Z*)-alkene after stereochemical assignment.

Let's try one last example to check you've fully understood. This one can be tricky so, again, take care. Consider how you would assign the configuration of the diene shown in Figure 2.41. Have a go at this example before you continue reading; you may be able to quickly assign the diene by eye.

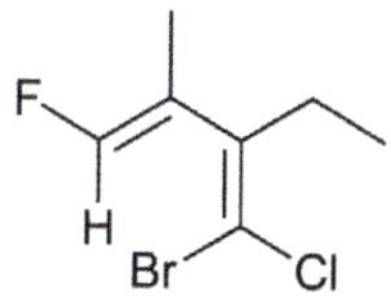

Figure 2.41   Exemplar diene for configurational assignment.

Let's start by breaking this molecule up; we will consider the left alkene (Figure 2.42).

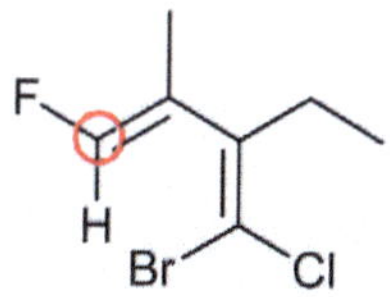

Figure 2.42   The carbon we are focusing on is circled.

On the carbon, the fluorine takes priority over the hydrogen. This gives us the assignment shown in Figure 2.43.

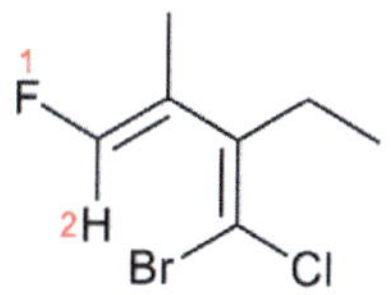

Figure 2.43   Diene with priorities so far.

The second carbon is trickier; the carbon is bonded to two other carbons (Figure 2.44).

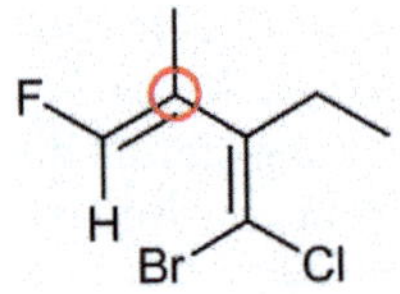

Figure 2.44   Diene, focusing on the circled carbon.

This means we need to go one more atom along to assess the priority of both carbons for the tiebreak. The methyl group on the carbon is only connected to hydrogens. The other has three bonds to carbons (Figure 2.45), therefore the atom connected to the three C atoms takes priority (Figure 2.46).

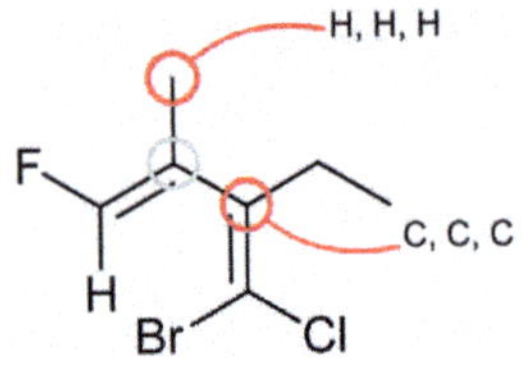

Figure 2.45   The carbon we are focusing on is circled in grey. The tied carbons are circled in red and have the bonded atoms labelled.

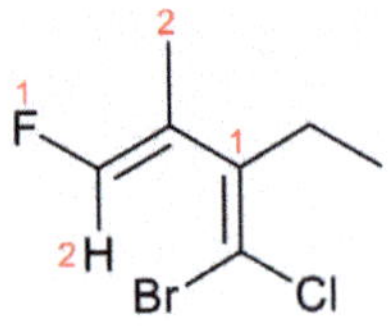

Figure 2.46   The left alkene is now fully labelled with priorities.

We will look at the bottom carbon now (Figure 2.47).

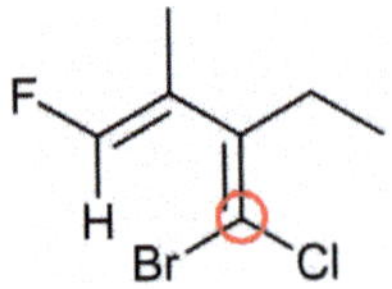

Figure 2.47   Diene with focus on the circled carbon.

This is pretty simple; the Br (atomic number = 35) takes priority over the Cl (atomic number = 17), as shown in Figure 2.48.

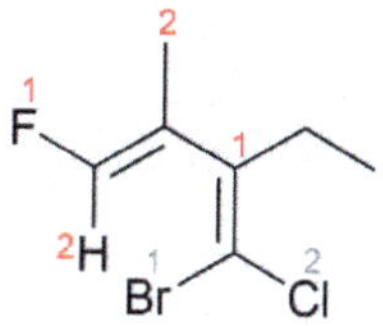

Figure 2.48   Labelled priorities for three of the carbons.

The last carbon is the toughest part of this assignment; again this carbon is bonded to two carbon atoms (Figure 2.49).

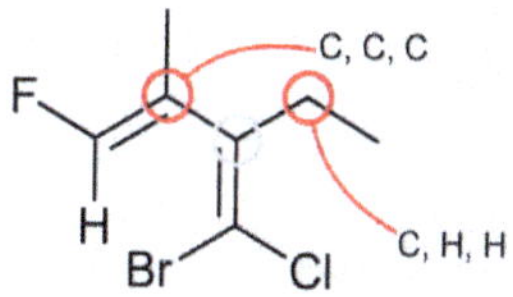

Figure 2.49   The carbon we are focusing on is circled in grey. The tied carbons are circled in red and have the bonded atoms labelled.

We need to work out which one takes priority! The right tied carbon is bonded to one carbon (and two hydrogens); the left tied carbon has three bonds to carbon atoms and therefore takes priority. Have a look at the assignment overview in Figure 2.50.

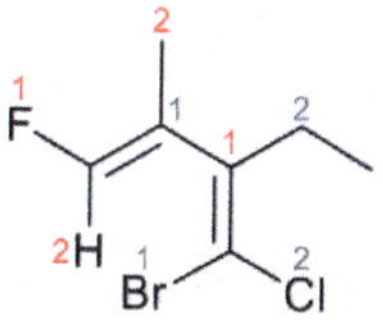

Figure 2.50   An overview of our assignment on the exemplar diene.

We can see that the priority groups on the left alkene are on opposite sides; this means it is assigned as *E*. The right alkene has its priority groups on the same side, so is labelled *Z*! Ultimately, we find the molecule is assigned as (1*E*,3*Z*)- (or fully as (1*E*,3*Z*)-3-(bromochloromethylene)-1-fluoro-2-methylpent-1-ene).

2.4   Structure Determination Using ¹H NMR Spectroscopy

NMR spectroscopy is a powerful tool for elucidating the unknown structure of a molecule, allowing chemists to see subtle structural features. In this section, we will discuss proton (¹H) NMR spectroscopy. Some prior knowledge of this technique is assumed but the most essential information will be provided. We will apply the technique to analysing alkenes in particular, using NMR observations

> **▮ Checkpoint**
>
> You should now know the Cahn–Ingold–Prelog (CIP) rules to assign the stereochemistry of alkenes.

to make informed suggestions about the configuration, starting with simple chains. We will restrict the discussion to 1D spectra as that is what most undergraduates will be most familiar with.

This spectroscopic technique is reasonably advanced but is the method that chemists often employ to confirm the stereochemistry of alkenes! You will come across how to inspect NMR spectra in the first or second year of an undergraduate course. If you haven't come across NMR spectroscopy, feel free to gloss over this section, but it is recommended even as a quick primer to the technique, as we will discuss some of the main concepts behind NMR.

2.4.1 ¹H NMR Spectroscopy Essentials

Nuclei have an inherent property called spin; proton nuclei have spin ½ (this is based in the constituency of protons, specifically the fundamental particles that they are composed of). This spin value essentially means that the proton behaves like a bar magnet in the presence of a magnetic field. When a field is applied, the bar magnets, which are exemplifying protons, will orient themselves parallel to the applied field. The bar magnet can sit antiparallel to the applied field; however, this is unlikely and is unstable! In NMR, radio waves are used to promote these bar magnets to their antiparallel states; chemists like to use the term chemical shift (δ) to describe the energy required to do this. Feel free to do some extra reading on NMR; it is a very interesting topic and very useful knowledge to a chemist!

Don't worry if this 'theory' feels a bit out of place here. One of the places where the stereochemistry of a molecule is obvious is in NMR and other spectra, so we're keen for you to make the link.

2.4.2 Shielding

What affects the chemical shift or frequency the protons resonate at? In theory, all isolated protons resonate at the same frequency; however in reality they are not isolated! In fact, the magnetic field they experience is influenced by the electrons around them, as the electrons themselves can create magnetic fields. On application of a magnetic field, the electrons can revolve around the proton, and as electrons are charged a magnetic field will be generated (Figure 2.51). This secondary magnetic field opposes the applied field (Figure 2.52).

As you might imagine, if there were a higher electron density around a proton, we might see the opposition of this field more. This is exactly the case! As we increase the electron density around the proton, we see that the secondary field gets stronger, and lower-energy radiation is needed to bring the nuclei into resonance. This effect is referred to as shielding. The higher the electron density, the more the proton is shielded. The lower the electron density,

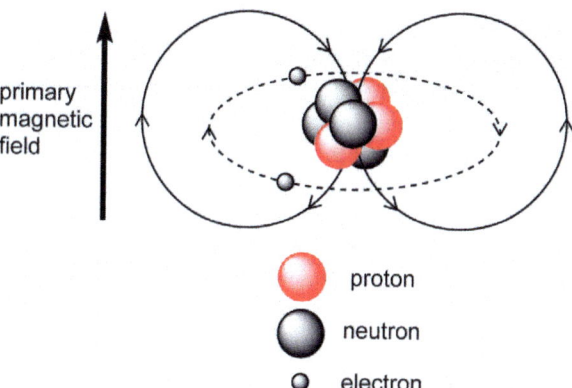

Figure 2.51   Electrons revolve around a proton in an applied magnetic field, generating a secondary field that opposes the applied field.

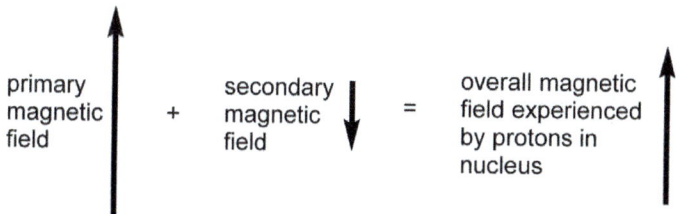

Figure 2.52   Visual representation of the impact of the opposing magnetic fields of Figure 2.51 interacting and the overall magnetic field experienced by the proton.

the lower the opposing field; this is referred to as deshielding. What causes an increase or decrease in electron density? Often, electropositive and electronegative substituents, respectively. For example, lithium is very electropositive and fluorine is very electronegative (Figure 2.53).

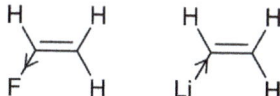

Figure 2.53   Lithium pushes electron density to the rest of the molecule, fluorine pulls electron density from the rest of the molecule.

The effect of this is that electron density will have a large impact on chemical shift, and we will see that it also affects the coupling between protons.

2.4.3 Coupling

If you have seen an NMR spectrum you will have noticed that the peaks can seem quite complex, sometimes forming patterns of peaks. This is caused by an effect called coupling and is due to the interactions between non-equivalent NMR-active nuclei. We have rationalised the spin of nuclei as tiny bar magnets. We have also discussed how the magnetic environment of the nuclei will affect their chemical shifts. The intuition that a bar magnet can affect another bar magnet's magnetic environment seems obvious, doesn't it? This interaction between the nuclei is transmitted by the electrons in the orbitals between the two nuclei. In alkenes, the three-bond coupling is of great importance. As the effect is transmitted through electrons, or electron density, the orbital overlap will affect the **coupling constant**, J. J denotes the coupling between protons; it is often prefaced by a superscript number denoting the number of bonds the coupling nuclei are away from one another, e.g. ^3J for alkene protons a and b (Figure 2.54), as you have to go along three bonds to get between the nuclei. You may also see J written as J_{ab} where a and b are labelled coupled nuclei, i.e. proton 'a' and proton 'b'. Coupling constants are measured in hertz. Coupling constants are often communicated using the Karplus equation (met in later chapters). The more the alkene protons are in the same plane, the stronger the coupling, as this is where orbital overlap is maximised (Figure 2.54).

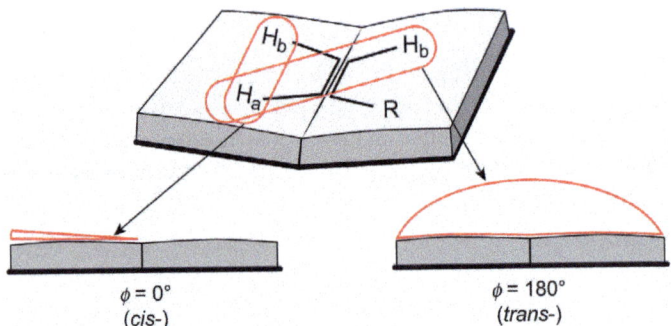

$\phi = 0°$
(cis-)

$\phi = 180°$
(trans-)

Figure 2.54   An exemplar alkene lying on the surface of a fully open book; the protons on the same page have $\phi = 0°$ and protons on adjacent pages have $\phi = 180°$.

Dihedral angle (ϕ) is a term chemists use to quantify the angle of two bonds originating from two atoms in a central bond. Consider comparing the compound to a book (Figure 2.54); two paragraphs on the same page (the same side of the spine of the book) have a dihedral angle of 0°. If we want ϕ for paragraphs on adjacent pages, we will find that $\phi = 180°$ (or less, assuming you haven't snapped the spine of the book!).

When we look at alkene coupling, we see a proton at $\phi = 0°$ and a proton at $\phi = 180°$ (in Figure 2.55, this would be H_a–H_b (cis) and H_a–H_b (trans), respectively). The orbital overlap for $\phi = 180°$ is ever so slightly better; this results in somewhat higher coupling constants in trans protons and somewhat lower coupling constants for cis protons (Figure 2.55).

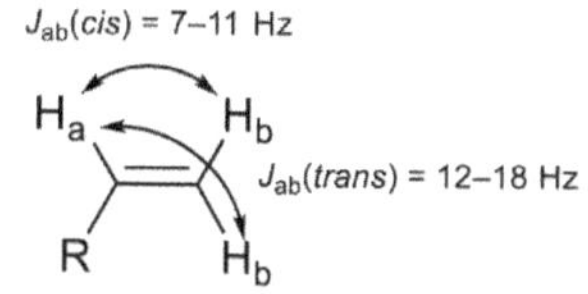

$J_{ab}(cis)$ = 7–11 Hz

H_a H_b

$J_{ab}(trans)$ = 12–18 Hz

R H_b

Figure 2.55 Typical coupling constants (written as J followed by the protons that are coupling) across an alkene. For cis protons, $\phi = 0°$, and for trans protons, $\phi = 180°$.

The last thing to note is the effect of substituents on the coupling constants in an alkene. We have discussed the effect of electron density already, so you may be able to suggest the effect of an electron-withdrawing group; it often lowers the magnitude of the coupling constants, while an electron-donating group often increases coupling constants (Figure 2.56) in alkenes. However, in each case you'll notice that the trans coupling is greater than the cis.

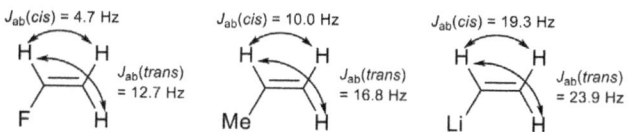

$J_{ab}(cis)$ = 4.7 Hz $J_{ab}(cis)$ = 10.0 Hz $J_{ab}(cis)$ = 19.3 Hz

H   H   H   H   H   H

$J_{ab}(trans)$ $J_{ab}(trans)$ $J_{ab}(trans)$
= 12.7 Hz   = 16.8 Hz   = 23.9 Hz

F   H   Me   H   Li   H

Figure 2.56 Coupling constants for a selection of alkenes with an electron-withdrawing group or electron-donating group.

Back in Section 2.1.2, we briefly mentioned amides and how they can also display stereochemistry, owing to the restricted C–N bond rotation. This can also be observed in NMR spectra and we'll come back to this in Section 4.6.

> **◾ Checkpoint**
>
> You should now be comfortable with how NMR can be used to elucidate the structure of an alkene.

2.5 Conclusion

Well done. You've made it to the end of Chapter 2!

At this point we hope you've gained a full understanding of why some alkenes exhibit stereochemistry and that different geometric isomers have different properties. With respect to terminology and conventions, you should feel confident with when and how to assign E/Z and *cis/trans* prefixes to describe an alkene as well as know the Cahn–Ingold–Prelog (CIP) rules. Finally, we hope you are comfortable with using NMR spectroscopy to elucidate the structure of an alkene.

The good news is that not only are you now an expert on alkene stereochemistry, but you have gained loads of useful knowledge to apply in the 3D world we are about to enter.

Exercises

Exercises (1) and (2) are straightforward recall questions to check you have the basic concepts and might form a simple opener in an exam question. Exercises (3) and (4) require you to apply stereochemical conventions to a series of alkenes and might be found in the next part of an exam question. Likewise, exercise (5) asks you to apply your basic NMR knowledge. Exercise (6) is more challenging, requiring you to bring your stereochemical and NMR knowledge together and might form a later part of an exam question. Good luck!

1. Draw the general structure of a *cis*-alkene and a *trans*-alkene.
2. *Trans*-fatty acids are present in foods like cakes and general 'comfort foods'. Elaidic acid (see Figure 2.57) is an example of a *trans*-fatty acid.

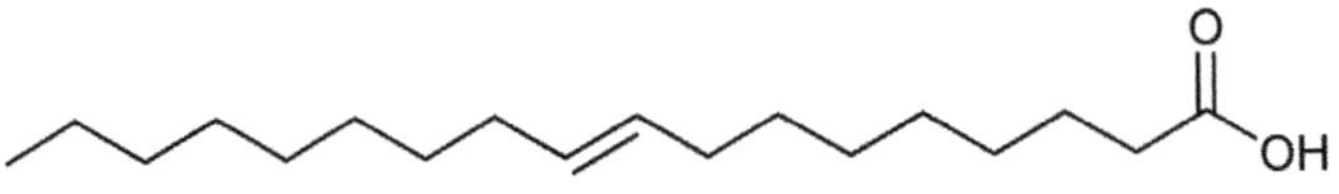

Figure 2.57 *Trans*-elaidic acid.

   Why don't *trans*-fatty acids convert to the *cis* form on standing? (Hint: Consider the bonding of the alkene).
3. Label the following molecules using *cis/trans* nomenclature:

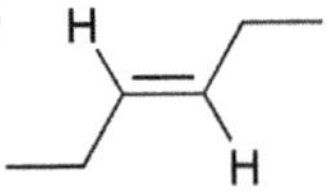

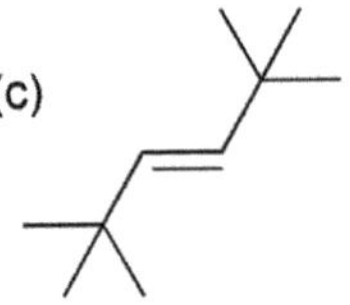

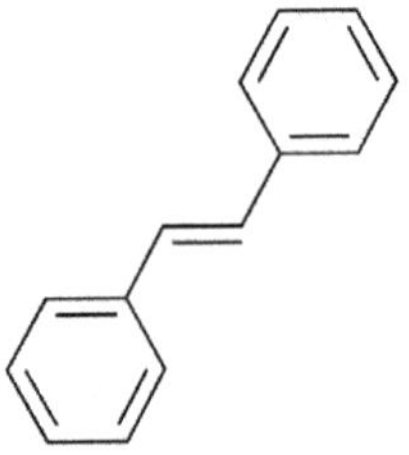

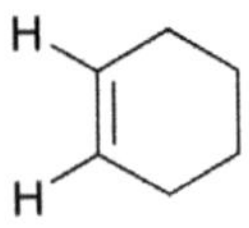

4. Label the following molecules using the *E/Z* nomenclature:

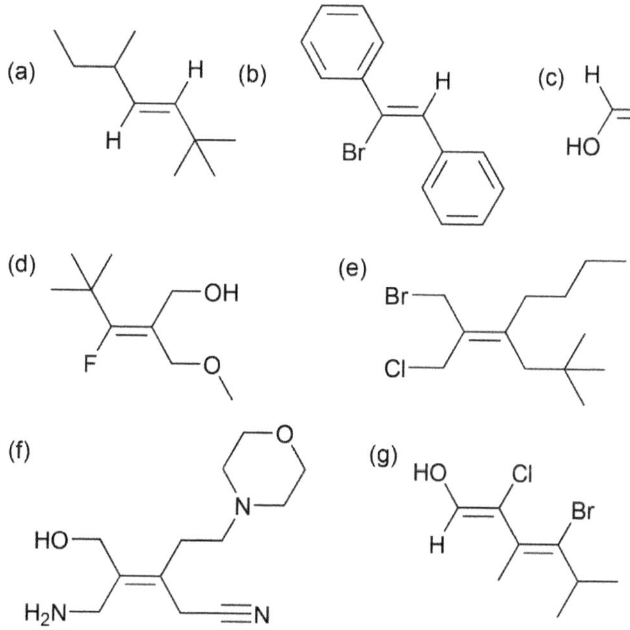

5. A ^{1}H NMR analysis shows the coupling between the alkene protons of pent-2-ene as 10.9 Hz. What configuration would you expect this to be? (Hint: Referring to Section 2.4.3 will definitely help here.)

6. Figure 2.58 shows a general monosubstituted alkene. What trend would you expect for the *trans* coupling values between the protons as R takes the value of the successive groups listed in the figure? Consider what effect on the electron density these groups would have. (Hint: They are in order of electronegativity.)

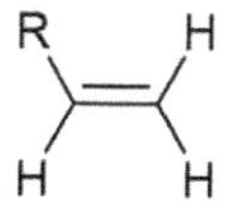

R = F, Cl, Br, I,
CH$_3$, Mg, Li

Figure 2.58 A general monosubstituted alkene with progressively less electronegative substituents bonded to it.

Answers

1.

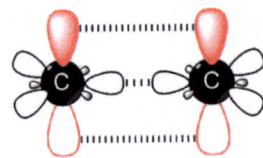

For the *cis*-alkene, the R substituents are on the same side. For the *trans*-alkene, the opposite is true.

2. The interconversion would require rotating around the C=C bond, which has restricted rotation. This is due to the bonding of an alkene, specifically due to the interaction of the p orbitals.

   We can see in Figure 2.59 that the p orbitals line up nicely and the overlap is favourable. As we rotate about the C=C bond, we reduce the overlap of the orbitals, eventually breaking the π bond. This means that it would require breaking the π bond to interconvert the isomers, meaning that a *cis*-alkene can't convert into a *trans*-alkene and *vice versa*.

Figure 2.59   Depiction of the bonding in an alkene, showing the σ bond and π bond.

3.

(a)   *cis-*

(b)   *trans-*

(c)   *trans-*

(d)   *trans-*

(e)   *cis-*

4.

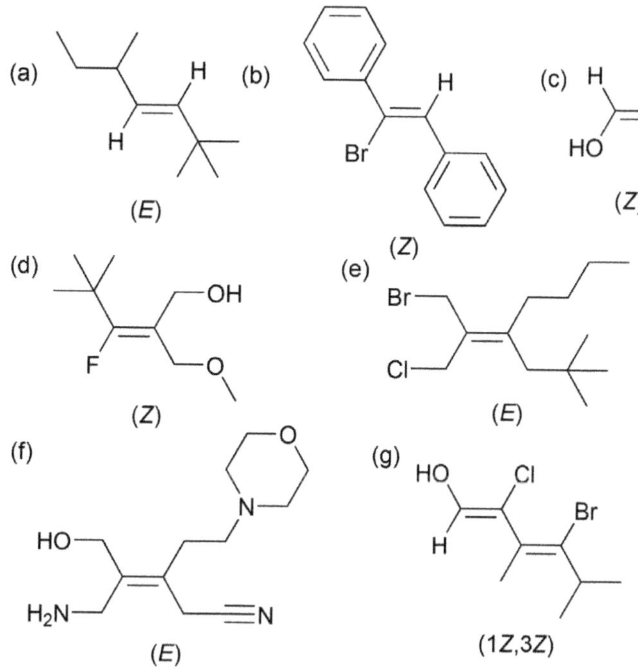

(a)

(E)

(b)

(Z)

(c)

H   Cl

HO   Br

(Z)

(d)

—OH

F   —O

(Z)

(e)

Br—

Cl—

(E)

(f)

HO—

H₂N—

≡N

(E)

(g)

HO   Cl

H   Br

(1Z,3Z)

5. We see, using the propene from Figure 2.56 as a model, that the *cis* coupling is lower. This is true here; the *cis* coupling is lower than the *trans* coupling. The *trans* coupling would be around 15 Hz.
6. We would expect the coupling to increase as the electronegativity of the R group decreases.

These values only apply for the monosubstituted alkene shown in Figure 2.58. Remember that there are lots of different effects that come into play with NMR; take care to consider them. These effects will impact the chemical shift and coupling values. In reality, many of these values are calculated by computers but we can make approximations using rudimentary concepts.

R group	J_{trans}/Hz
F	12.7
Cl	14.6
Br	15.2
I	15.9
CH_3	16.8
Mg	23.0
Li	23.9

By the End of This Chapter You Will:

- [ ] Understand what makes molecules chiral or achiral.
- [ ] Be able to assign the configuration of a stereogenic centre as *R* or *S*.
- [ ] Understand the differences in the properties of enantiomers and diastereoisomers.
- [ ] Understand the definition of pairs of enantiomers, pairs of diastereoisomers, racemates (racemic mixtures) and *meso*-compounds.
- [ ] Understand how we measure the optical rotation of compounds.

What You Will Get from This Chapter

In this chapter we will look initially at simple molecules containing one stereogenic centre, how we name these molecules and how we define the arrangement of the groups in 3-dimensions. We will then take a similar approach considering slightly more complicated examples with two or more stereogenic centres.

Stereogenic Centres, Enantiomers and Diastereoisomers

In Chapter 1, we saw that stereochemistry is extremely important, particularly because different arrangements of atoms or substituents in three dimensions can lead to a very different biological response, as in the example of thalidomide. It is therefore essential that we have systematic methods to name and define different types of stereoisomer. In this chapter, we will look initially at simple molecules containing one **stereogenic centre**, how we name these molecules and how we define the arrangement of the groups in three dimensions. We will then take a similar approach in considering slightly more complicated examples with two or more stereogenic centres.

3.1 What Makes Something Chiral?

Consider your hands (Figure 3.1); they are generally pretty close to being mirror images of each other. If you attempt to superimpose (position an object exactly on another object) one of your hands on the other one, you find there is no way to achieve it, no matter how you rotate them (*n.b.* placing them palm to palm is not superimposing them). This is what is known as chirality: *a **chiral** object is one that is non-superimposable on its mirror image.*

mirror plane

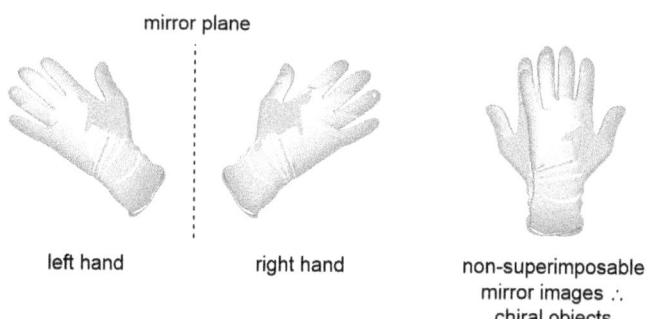

left hand   right hand   non-superimposable mirror images ∴ chiral objects

Figure 3.1 Chiral macro-objects.

Let's take another macroscopic object, a standard wine glass (Figure 3.2), and consider its mirror image; hypothetically two of these wine glasses would be superimposable and so are not chiral *i.e.* **achiral.** You can also consider achiral objects as those having

We have a single top tip before you dive into Chapter 3. Get yourself a molecular model kit and use it! Evidence suggests that you can't really grasp stereochemistry until you've built your own models, picked them up, turned them round and really seen the three dimensions we are studying. If a standard model kit isn't right for you, there are some good new virtual tools, but you need to be able to manipulate the molecules *yourself* to gain this new understanding. Oh, and a small mirror may come in handy!

The ∴ symbol is a useful shorthand for 'therefore' which you will see used in many of the figures in Chapter 3.

an internal mirror plane: the wine glass in Figure 3.2 is symmetrical about a plane vertically down the middle.

Figure 3.2   Achiral macro-objects. Wine glass image from Wikipedia commons, used with permission. https://commons.wikimedia.org/wiki/File:Wine_Glass_(Red).svg.

> **❶ Key Learning Point**
>
> In this thought experiment, we can see that two wine glass images can occupy exactly the same space (are superimposable), whereas left and right hands cannot.

The same macroscale principle applies on a molecular level. For example, if we take the molecule bromochlorofluoromethane (Figure 3.3), starting with the isomer shown in red, taking its mirror image shown in black and attempting to superimpose the two (requires rotation to put the bromine and chlorine atoms on top of one another), we find that this is unachievable and therefore we have chiral molecules. We refer to molecules with this relationship as a pair of **enantiomers**. Have a go at building molecular models of the two enantiomers of bromochlorofluoromethane (shown in Figure 3.3) and convince yourself that they are non-superimposable. You'll use them later in the chapter too.

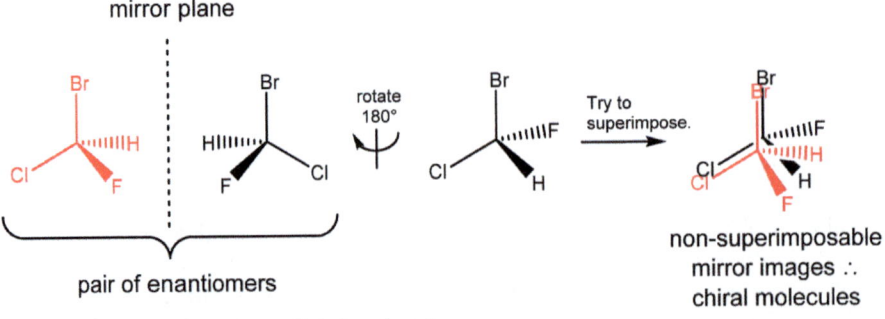

Figure 3.3   Chiral molecules.

In general, any sp^3 hybridised carbon atom (see Chapter 2) with four different substituents will give rise to chiral molecules and so we refer to the central carbon atom as being **stereogenic**.

> ❗ **Key Learning Point**
>
> A pair of enantiomers are non-superimposable mirror images of one another.

Remember to use the terms "stereogenic" when you are describing an atom and "chiral" for a molecule. Only if they are of course!

Now let's instead consider the molecule bromochloromethane (Figure 3.4), starting with the isomer shown in red and its mirror image shown in black. Have a go at superimposing the two – it is possible in this case.

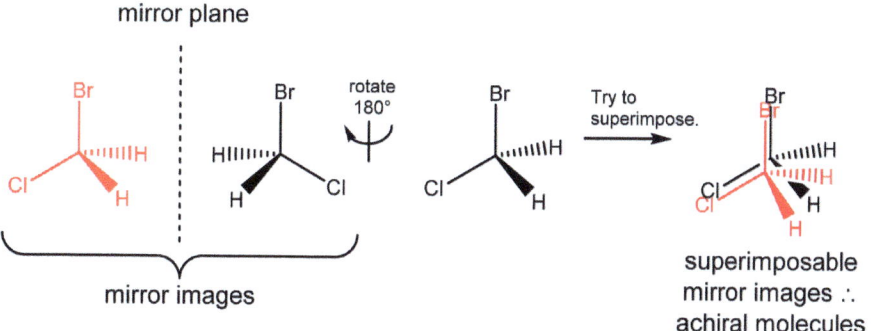

Figure 3.4 Achiral molecules.

Yes, that's right, we find that this is now achievable and therefore we have achiral molecules. Indeed the molecules shown in red and black are the same species. Just as with the wine glass, we can also consider molecules as being achiral if they possess an internal plane of symmetry, in this case, the plane of the paper (*i.e.* imagine turning the page to upright and looking straight down it towards the centre of the book. This is the mirror plane for this molecule.). It really will help to build molecular models of the two depictions of bromochloromethane and convince yourself that they are superimposable.

You may be wondering what the fourth substituent is in the chiral sulfoxide (Figure 3.5). Pyrimidal sp³ hybridised centres with lone pairs can be considered to have four substituents. For simple, unconstrained amines, the two possible configurations can interconvert readily, by rapid **inversion** through a planar sp² hybridised transition state. For sulfoxides, **inversion** is typically much slower and enantiomers can be isolated if the R groups are different.

In general, any sp³ hybridised carbon atom bearing two identical substituents gives rise to achiral molecules, unless there is a stereogenic atom elsewhere in the molecule.

It is important to note that carbon is not the only atom that can be stereogenic. Other common stereogenic atoms include phosphorus, sulfur, nitrogen and many metals (see Chapter 7). Some examples of these stereogenic atoms in chiral molecules, bound to four different substituents, are shown in Figure 3.5.

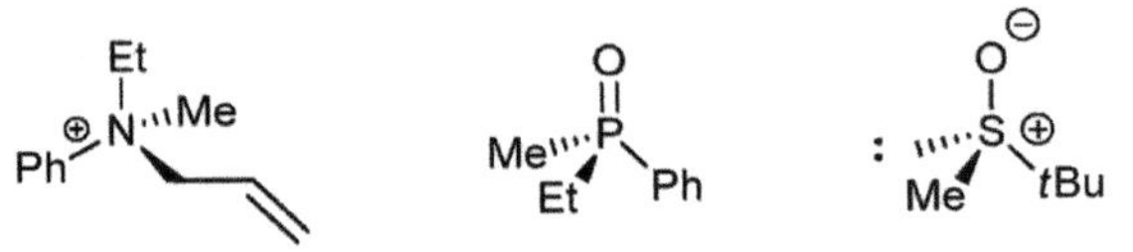

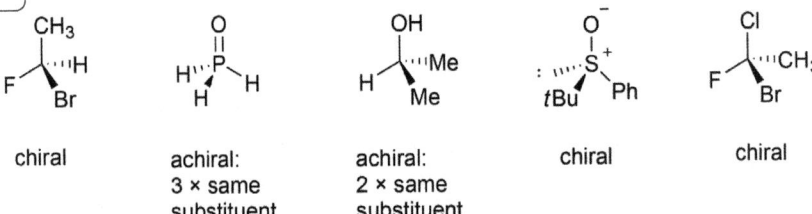

an ammonium ion   a phosphine oxide   a sulfoxide

Figure 3.5   Other stereogenic atoms.

Let's just check our understanding. Are the five molecules in Figure 3.6 chiral or achiral?

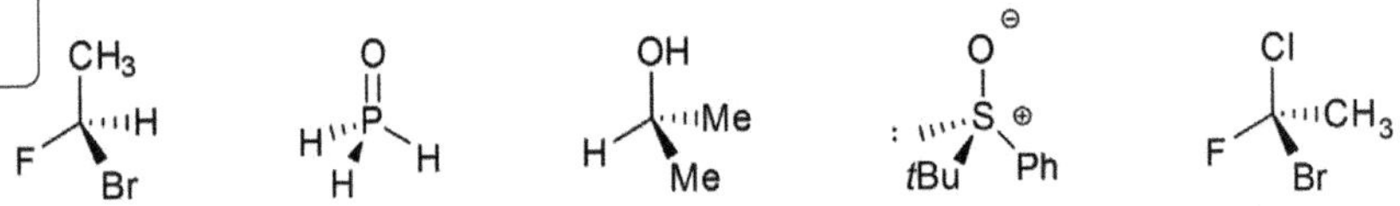

Figure 3.6   Checkpoint: can you determine whether a molecule is chiral or achiral?

◼ Checkpoint

You should now understand what makes molecules chiral or achiral.

Hopefully, you spotted that the second and third structures both have at least two identical substituents, so are not chiral. The others are chiral (Figure 3.7).

chiral   achiral: achiral: chiral   chiral
                3 × same   2 × same
                substituent   substituent

Figure 3.7   Molecules with two or more identical substituents are achiral.

❶ Key Learning Point

The configuration is the spatial arrangement of atoms in a molecule that can only be altered by breaking and making bonds.

3.2 How to Define Enantiomers' Configuration

In Chapter 2, we met the Cahn–Ingold–Prelog (CIP) rules for defining the stereochemistry of alkenes. The same rules can be applied to assign the configuration of a stereogenic centre as being either *R* (from the Latin *rectus*, meaning right-handed) or *S* (from the Latin *sinister*, meaning left-handed). If you want some more practice at stereochemical *cis/trans* and *E/Z* assignment go back to Section 2.3.

The CIP rules for assigning a stereogenic atom as *R/S* are as follows:

1. Prioritise the four substituents immediately attached to the **stereogenic centre** (1 bond away) with numbers 1 (highest priority) to 4 (lowest priority) based on atomic number. Atoms with a higher atomic number get higher priority.
2. Arrange the molecule such that the lowest-priority substituent (priority 4, with the lowest atomic number, typically hydrogen) is pointing away from you.
3. Look at the other three substituents. If moving from priority 1 → 2 → 3 proceeds in a clockwise fashion, the molecule is assigned the *R* configuration. If 1 → 2 → 3 proceeds in an anticlockwise fashion, it is assigned as *S*.

We will have a go at assigning the absolute configuration of the two enantiomers of bromochlorofluoromethane that we made models for earlier, shown again in Figure 3.8.

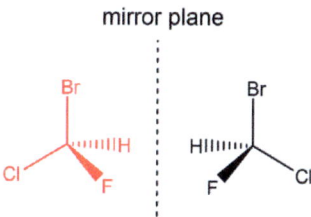

Figure 3.8   Enantiomers of bromochlorofluoromethane.

> ❗ **Key Learning Point**
>
> Absolute stereochemistry refers to the precise spatial arrangement of substituents around a stereogenic centre.

We will first consider the enantiomer shown in red.

Step 1. Prioritise the four substituents immediately attached to the stereogenic centre (1 bond away) with numbers 1 (highest priority) to 4 (lowest priority) based on atomic number. Atoms with a higher atomic number get higher priority.

Looking at Figure 3.9, of the four substituents (Br, Cl, F and H), bromine has the highest atomic number (35) and so takes priority 1. Chlorine has the next highest atomic number (17) and so takes priority 2. Fluorine has the next highest atomic number (9) and so takes priority 3. Hydrogen has the lowest atomic number (1) and so takes priority 4.

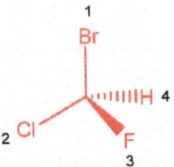

Figure 3.9   Assigning priorities through the CIP rules.

Step 2. Arrange the molecule such that the lowest-priority substituent (priority 4, with the lowest atomic number) is pointing away from you.

With your model kit, rotate the molecule of bromochlorofluoromethane so that you are looking down the C–H bond towards the hydrogen atom, which is our priority 4 substituent. Redraw the molecule with the priority 1–3 substituents as shown in Figure 3.10.

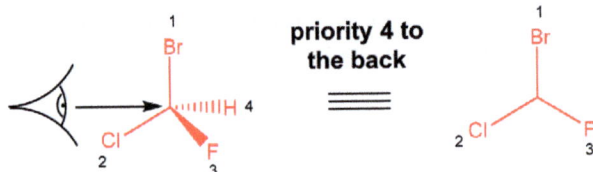

Figure 3.10   Arranging the molecule to have the atom with priority 4 at the back.

Step 3. Look at the other three substituents. If moving from priority 1 → 2 → 3 proceeds in a clockwise fashion, the molecule is assigned the *R* configuration. If 1 → 2 → 3 proceeds in an anticlockwise fashion, it is assigned as *S*.

Looking at Figure 3.11, moving from the priority 1 substituent (Br) to priority 2 (Cl) to priority 3 (F), we are going in an anticlockwise fashion and so we have the *S* enantiomer.

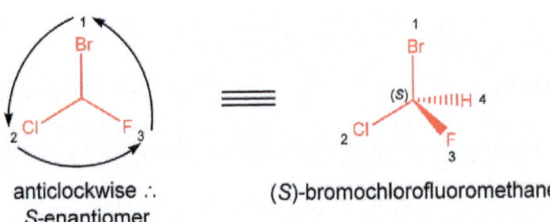

Figure 3.11   Assigning the stereocentre as *R* or *S*.

Let's now take the other enantiomer of bromochlorofluorometh-ane (shown in black in Figure 3.8) and apply the same rules.

Step 1. Prioritise the four substituents immediately attached to the stereogenic centre (1 bond away) with numbers 1 (highest priority) to 4 (lowest priority) based on atomic number. As before, atoms with a higher atomic number get higher priority (Figure 3.12).

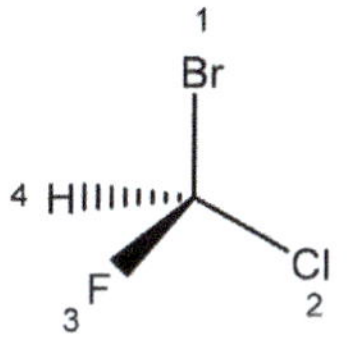

Figure 3.12 Assigning priorities through the CIP rules.

Step 2. Arrange the molecule such that the lowest-priority substit-uent (priority 4, with the lowest atomic number) is pointing away from you (Figure 3.13).

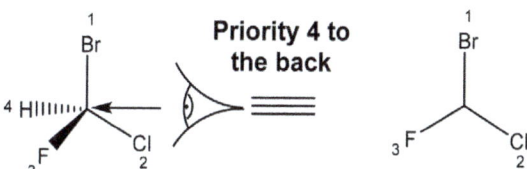

Figure 3.13 Arranging the molecule to have priority 4 to the back.

Step 3. Look at the other three substituents. Moving from the prior-ity 1 substituent (Br) to priority 2 (Cl) to priority 3 (F), we are going in a clockwise fashion and so we have the *R* enantiomer (Figure 3.14).

Not remembering to have priority 4 to the back is a really common mistake in exams.

---

❶ Key Learning Point

With priority 4 to the back, a clockwise arrangement of the three highest-priority substituents is the *R* configuration. Anticlockwise is *S*.

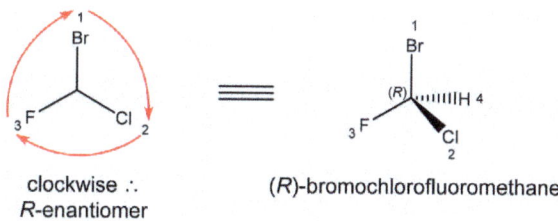

clockwise ∴
R-enantiomer

(*R*)-bromochlorofluoromethane

Figure 3.14 Assigning the stereocentre as *R* or *S*.

As you get more familiar with the method, you can eliminate the redrawing part of Step 2, as long as you ensure that the priority 4 substituent is pointing to the back (hashed bond).

Now we've got to grips with this, let's check and assign the configuration of the stereocentre in each of the three molecules in Figure 3.15 as either R or S, using the CIP rules.

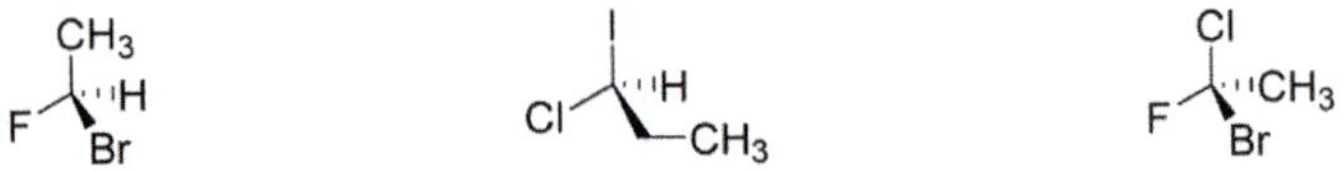

Figure 3.15   Can you assign a stereogenic atom as R or S?

Did you get R, S, S respectively (Figure 3.16)? If not, don't worry. Reread the rules, try to see where you went wrong and have another go.

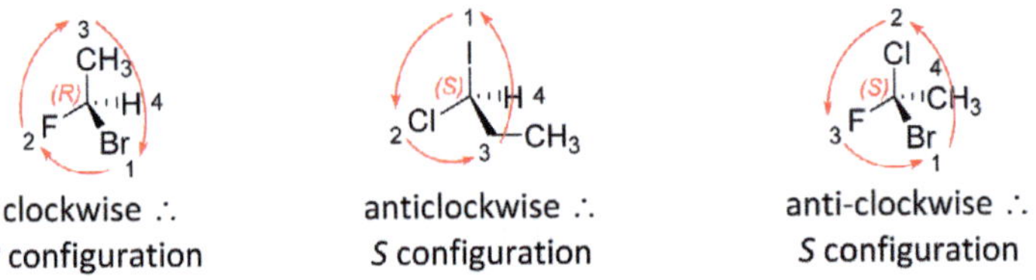

clockwise ∴
R configuration

anticlockwise ∴
S configuration

anti-clockwise ∴
S configuration

Figure 3.16   Using CIP rules to assign a stereogenic atom as R or S.

We can now assign a stereogenic atom as either R or S for simple chiral molecules in a convenient orientation, but not everything is so straightforward. Let's make a model of the enantiomer of 1-bromo-1-chloro-1-fluoropropane shown in Figure 3.17.

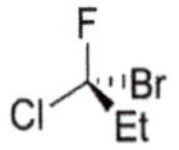

Figure 3.17   1-Bromo-1-chloro-1-fluoropropane.

Applying the CIP rules for the molecule (Figure 3.18), from Step 1 we find that Br takes priority 1, Cl priority 2 and F priority 3, while the carbon atom of the ethyl group has the lowest atomic number (6) and so takes priority 4. But for Step 2, the priority 4 substituent (Et) is pointing towards us and we need it to be pointing away from us. You need to rotate your model from the initial orientation until the ethyl group is pointing to the back. Look down the C–C bond and redraw the molecule depicting only the substituents with priorities 1–3. Now we can apply our rule for Step 3 and look from priority 1 → 2 → 3 and we find that we are moving in an anticlockwise fashion and so have the S enantiomer. We therefore have (S)-1-bromo-1-chloro-1-fluoropropane.

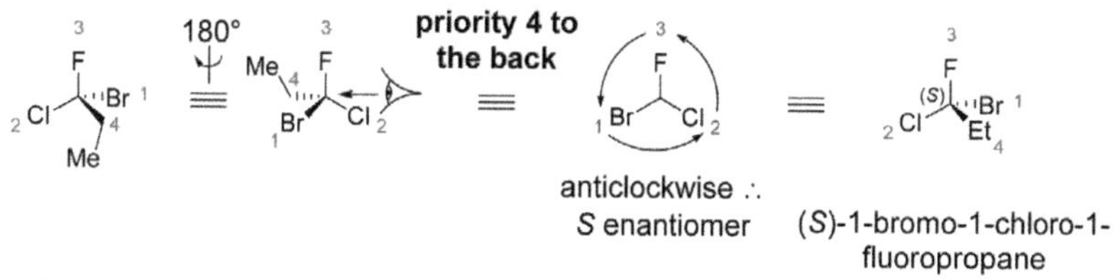

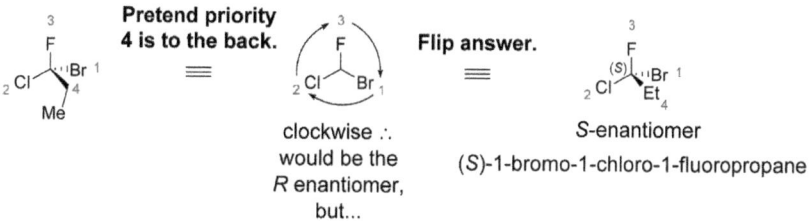

Figure 3.18 Assigning (S)-1-bromo-1-chloro-1-fluoropropane.

You may not always have access to your model kit, particularly in exams, and rotating molecules in your head to get the priority 4 substituent to the back frequently leads to mistakes, even for experienced chemists!

A quicker and arguably less error-prone method is to prioritise the substituents as normal, but then pretend that the priority 4 substituent is pointing to the back, the '**flip method**'. Look from priority 1 → 2 → 3, decide if it is clockwise or anticlockwise and hence *R* or *S*, respectively. The last step now should to be to flip your answer (*e.g. S* would be flipped to become *R*) to compensate for the priority 4 substituent not being pointed to the back. This method applied to the same enantiomer of 1-bromo-1-chloro-1-fluoropropane is shown in Figure 3.19.

Figure 3.19 Alternative flip method for when priority 4 is not at the back.

Another approach that works if priority 4 is not pointing to the back (particularly useful if it's drawn in the plane of the paper) is the **swap method**. It works on the principle that for swapping any two groups:

- An **odd** number of swaps will give the **opposite** configuration at that stereogenic centre compared with the original (not usually what we want).
- An **even** number of swaps will give the **same** configuration at that stereogenic centre compared with the original, which means that if we now assign the configuration it will be correct.

Let's apply this to the same example molecule, which we already know is (*S*)-1-bromo-1-chloro-1-fluoropropane (to show the method works).

Step 1. Assign the priorities 1 to 4 as usual.

If you are short of time in an exam, you might consider just swapping the priority numbers rather than redrawing the molecule each time you make a swap.

Step 2. Swap the priority 4 substituent with the substituent pointing to the back. **Note** that this gives the opposite configuration (useful if you want to draw the enantiomer or a **diastereoisomer** – see Section 3.6 – for any reason). Here we have swapped substituents 1 and 4.

Step 3. Leaving the priority 4 substituent alone (because we have it where we want it!), make any second swap to give an overall even number of swaps. Here we have swapped substituents 1 and 3.

Step 4. Proceed as usual, looking down towards priority 4, deciding if priority 1 → 2 → 3 is clockwise or anticlockwise and hence *R* or *S*, respectively. Here it is anticlockwise and so we have (*S*)-1-bromo-1-chloro-1-fluoropropane.

(Hint: This method is particularly useful if priority 4 is in the plane of the paper.) Figure 3.20 shows how the swap method works.

Don't get stuck with these different methods. Just find one that works for you! The flip and swap methods are great for avoiding simple mistakes in exams!

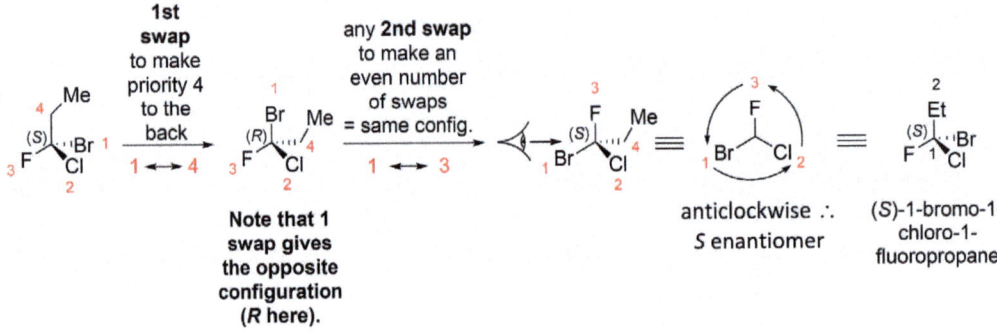

Figure 3.20   The swap method for assigning a stereogenic centre as *R* or *S*.

Try your hand at assigning the configuration of the stereocentre in the three molecules in Figure 3.21 as either *R* or *S*, using the CIP rules.

⚑ Figure 3.21   Checkpoint examples: easy examples with simple prioritisation.

Did you get *S*, *S* and *R*, respectively? Well done if you did! You can now assign stereogenic centres as *R* or *S* whatever the initial orientation of the molecule. If not, try again with the help of your molecular models and Figure 3.22.

The examples we have considered so far have all had four different atoms attached to the stereogenic atom. What happens if we have two or more of the immediately attached atoms the same? In

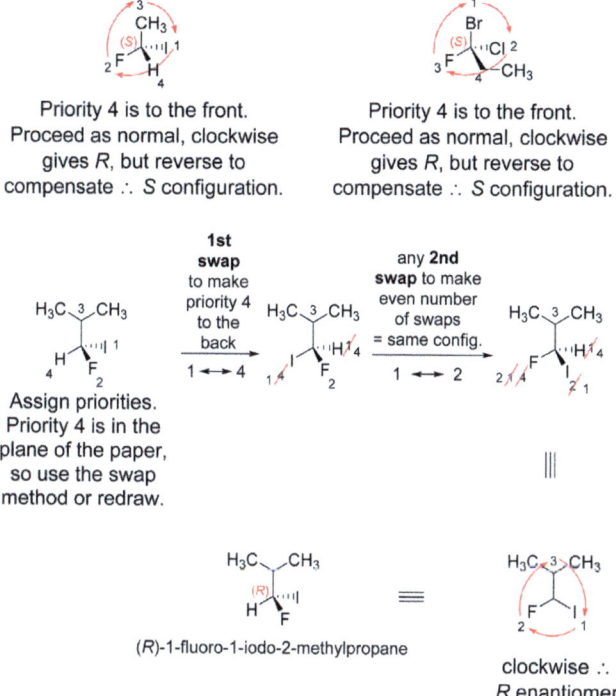

Figure 3.22   Examples with relatively simple prioritisation.

Chapter 2, you encountered a strategy to deal with this situation for prioritising substituents; when you have the same atom, move one bond further away, to the next set of atoms (*i.e.* two bonds from the stereogenic centre), and consider the atomic number of these atoms. More bonds to higher atomic number atoms takes priority (for these purposes, we consider double or triple bonds as two or three bonds to the same atom, respectively).

> ❗ **Key Learning Point**
>
> A bond to a higher number atom gives that atom priority. If there's a tie, more bonds to higher number atoms (counting multiple bonds multiple times) takes priority.

Let's make a model of the enantiomer of 2-iodobutane shown in Figure 3.23.

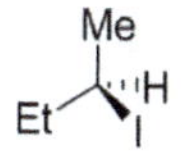

Figure 3.23   2-Iodobutane.

Looking at Figure 3.24, priorities 1 and 4 are simple; I (iodine) has the highest atomic number (53) so takes priority 1 and H has the lowest atomic number (1) so takes priority 4. The other atoms one bond from the stereogenic carbon are both carbon and so we now need to look one bond further away to distinguish between them. The carbon of the methyl group has $3 \times H$ (atomic number 1) attached, while the first carbon of the ethyl group has $1 \times C$ (atomic number 6) and $2 \times H$ (atomic number 1) attached. The carbon with more bonds to higher atomic number atoms takes priority, so the ethyl group takes priority 2 and the methyl group priority 3. We can now carry on as before; rotating the molecule to look down the C–H (priority 4) bond and looking at priority $1 \rightarrow 2 \rightarrow 3$, we find we are moving in a clockwise direction and so we have the R enantiomer, so (R)-2-iodobutane.

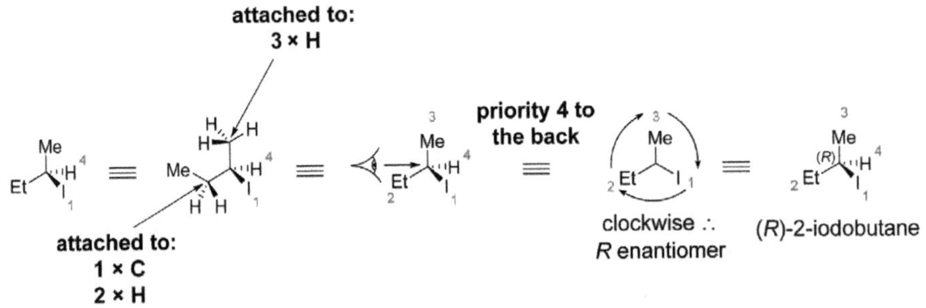

Figure 3.24   Assigning the **absolute stereochemistry** when two atoms are the same.

If we consider the biological building blocks amino acids (Figure 3.25), we find that they all (except in the case of glycine, where R = H) contain a stereogenic carbon atom and are therefore chiral molecules.

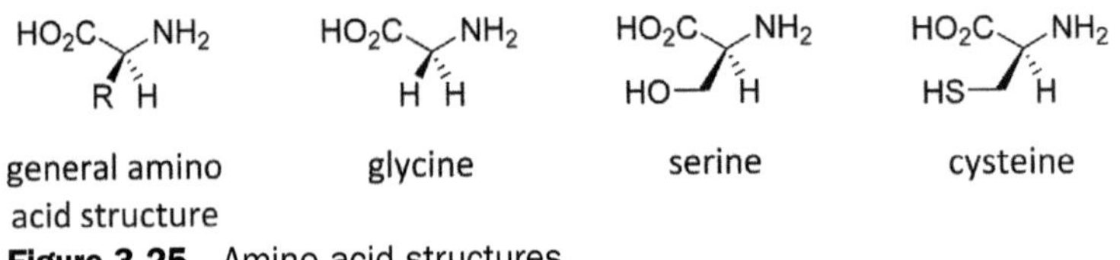

Figure 3.25   Amino acid structures.

Now make a model of the amino acid serine, shown in Figure 3.26, making sure you have the same enantiomer as depicted. If we assign the configuration of the stereogenic carbon using the CIP rules, from Step 1, we find that N has the highest atomic number (7) so takes priority 1 and H has the lowest atomic number (1) so

Writing out, for example, $1 \times C$, $2 \times H$, as in Figure 3.24, is good practice. It avoids mistakes and communicates your working to your teacher!

takes priority 4. The other atoms one bond from the stereogenic centre are both carbon and so we now need to look one bond further away to distinguish between them. The carbon of the carboxylic acid group has $3 \times O$ (atomic number 8) attached while the carbon of the alcohol group has $1 \times O$ (atomic number 8) and $2 \times H$ (atomic number 1) attached. More bonds to higher atomic number atoms takes priority, so the carboxylic acid group takes priority 2 and the alcohol group priority 3. We can now carry on as before; rotating the molecule to look down the C–H (priority 4) bond (Step 2) and looking at priority $1 \rightarrow 2 \rightarrow 3$ (Step 3), we find we are moving in an anticlockwise direction and so we have the S enantiomer, (S)-serine.

> Remember, in the CIP rules multiple bonds count multiple times.

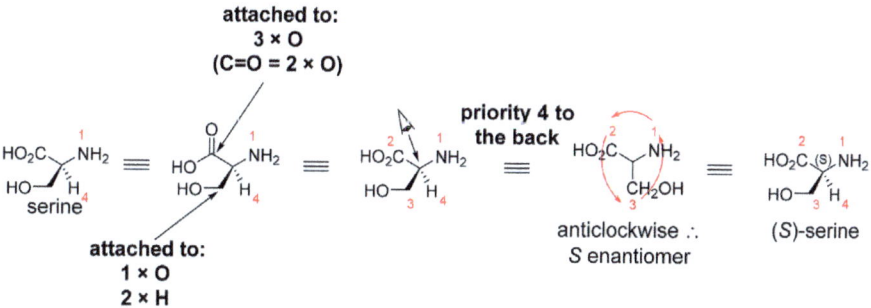

Figure 3.26   Assigning the absolute stereochemistry of amino acid derivatives.

All naturally occurring amino acids have S-configured stereogenic carbons with two exceptions: glycine (R = H), which has no stereogenic carbon, and cysteine. To solve this mystery, please make a molecular model of cysteine (see Figure 3.25), making sure you make the same enantiomer as that depicted, and assign the CIP priorities to the four groups. Maybe things will be getting clearer!

Similarly to serine, if we assign the configuration of the stereogenic carbon using the CIP rules, from Step 1, we find that N has the highest atomic number (7) so takes priority 1 and H has the lowest atomic number (1) so takes priority 4. The other atoms one bond from the stereogenic centre are both carbon and so we now need to look one bond further away to distinguish between them. The carbon of the carboxylic acid group has $3 \times O$ (atomic number 8) attached while the carbon of the thiol group has $1 \times S$ (atomic number 16) and $2 \times H$ (atomic

> Don't make the common mistake of prioritising CO_2H over NH_2 when using CIP rules with amino acids. The N is directly attached to the stereogenic carbon, so is considered before the oxygens.

Just to recap, a single attached atom of higher atomic number is all that is needed for that group to win the 'tiebreak'. This is a really common mistake in exams.

number 1) attached. Any bond to a higher atomic number atom takes priority (*i.e.* in this case one sulfur beats three oxygens, similar to a trump card winning in the card game whist), so the thiol group takes priority 2 and the carboxylic acid group priority 3. We can now carry on as before; rotating the molecule to look down the C–H (priority 4) bond (Step 2) and looking at priority 1 → 2 → 3 (Step 3), we find we are moving in a clockwise direction and so we have the *R* enantiomer, (*R*)-cysteine (Figure 3.27).

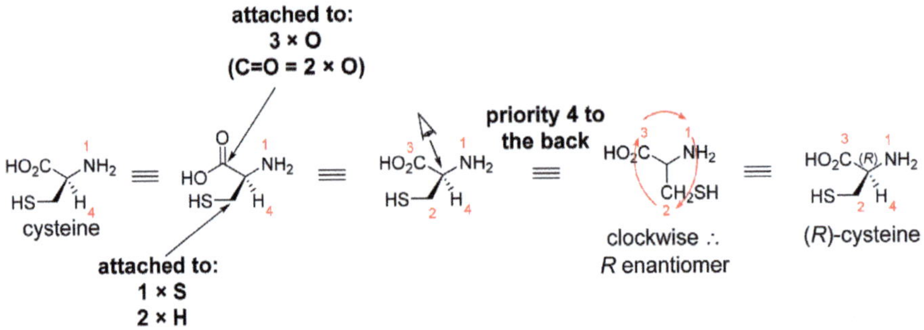

Figure 3.27   Assigning the absolute stereochemistry of cysteine derivatives.

We've rightly told you to use atomic number, not atomic mass, to assign CIP priorities. But when you have different isotopes of the same element, atomic mass is all that distinguishes them.

If we take the achiral amino acid glycine and replace one of the hydrogen atoms with a deuterium atom, we now have four different substituents around the carbon and hence a stereogenic centre (Figure 3.28). In instances like this, where two of the substituents are isotopes with the same atomic number (H and D both have the atomic number 1), we now use atomic mass as a 'tiebreaker' to differentiate between them. Hydrogen has an atomic mass of 1 and deuterium has an atomic mass of 2, so deuterium takes priority 3 and hydrogen priority 4. We can now carry on as before; rotating the molecule to look down the C–H (priority 4) bond (Step 2) and looking at priority 1 → 2 → 3 (Step 3), we find we are moving in an anticlockwise direction and so we have the *S* enantiomer, (*S*)-2-deuteroglycine.

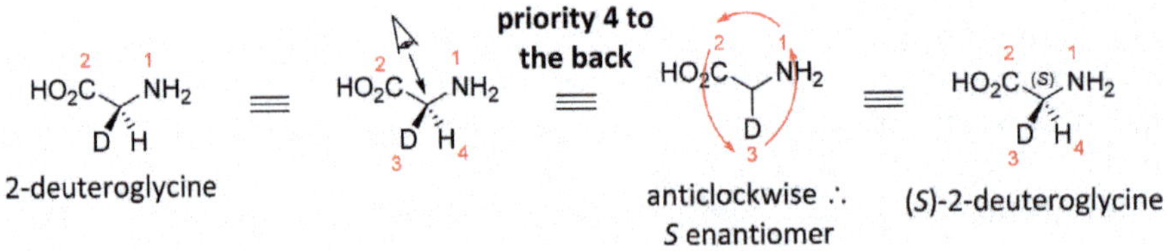

Figure 3.28   Using atomic mass as a priority 'tiebreaker' for isotopes.

Let's stop at a final CIP checkpoint and make sure that you can assign the configuration of the stereocentre in the molecules in Figure 3.29 as either R or S, using the CIP rules.

Figure 3.29   Assigning absolute configuration in more difficult examples.

Hopefully you got S, S, R, S, respectively (Figure 3.30). If not, check you used the tiebreak rules correctly and remember to count a double bond as two single bonds.

> **■ Checkpoint**
>
> You should now be able to assign the configuration of a stereogenic centre as R or S.

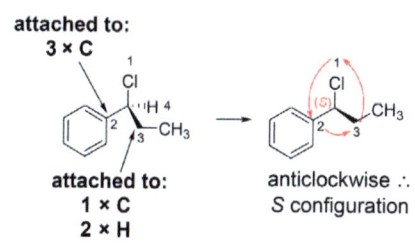

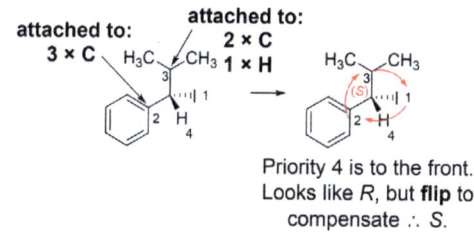

Figure 3.30   Assigning configuration can require care with prioritisation.

<table>
<tr><td>**3.3**</td><td>**Properties of Enantiomers and Other Stereochemical Descriptors**</td></tr>
</table>

Pairs of enantiomers exhibit identical physical and spectroscopic properties (boiling point, density, NMR spectra *etc.*) with one exception: *they rotate the plane of polarised light in opposite directions,* clockwise, denoted (+) or *d* (from the Latin *dexter*, meaning 'to the right') or anticlockwise denoted (−) or *l* (from the Latin *laevus*, meaning 'to the left'). We can measure this property, known as *optical rotation*, using an instrument known as a polarimeter (Figure 3.31). A monochromatic light source (typically a sodium or mercury lamp) is polarised (the light waves oscillate in one plane *e.g.* vertically up and down) and directed through an optical cell containing the sample to be analysed. If the sample contains an excess of one enantiomer of a compound, the light will be rotated through an angle, *α*, which can be detected and measured. The value of *α* is affected by temperature, sample concentration and sample path length (as well as the wavelength of light used) and so the angle, *α*, is standardised by converting to a specific rotation [*α*] to account for this, using the formula

$$[\alpha]_{\lambda}^{T} = \frac{\alpha}{l \times c}$$

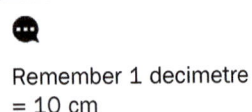

Remember 1 decimetre = 10 cm

where *T* is the temperature the experiment is performed at, *λ* is the specific wavelength of monochromatic light used in the lamp (typically, the sodium D line of 589.3 nm), *α* is the angle of rotation (in degrees), *l* is the path length of the sample (in decimetres) and *c* is the concentration of the sample solution (in grams per millilitre, quoted as neat (no solvent) or with the solvent specified).

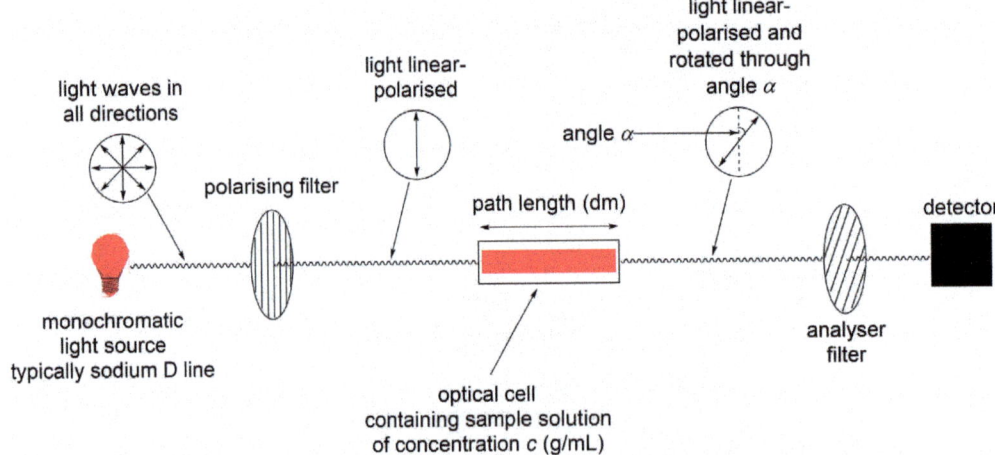

light waves in all directions

monochromatic light source typically sodium D line

polarising filter

light linear-polarised

path length (dm)

optical cell containing sample solution of concentration *c* (g/mL)

angle *α*

light linear-polarised and rotated through angle *α*

analyser filter

detector

Figure 3.31   A polarimeter set up, for measuring optical rotation.

If we consider a specific example, butan-2-ol, and look at the *R* and *S* enantiomers individually (Figure 3.32), we find that for the neat liquids at 24 °C, the *R* enantiomer has a specific rotation of −13.74 and the *S* enantiomer has a specific rotation of +13.80, so we can conclude that (within experimental error) a pair of enantiomers will have *equal magnitude but opposite specific rotations.*

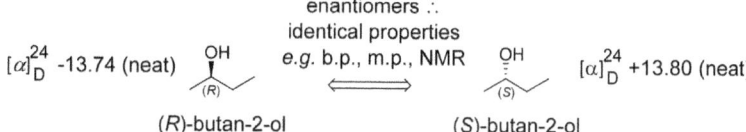

$[\alpha]_D^{24}$ -13.74 (neat) $[\alpha]_D^{24}$ +13.80 (neat)

(*R*)-butan-2-ol   (*S*)-butan-2-ol

Figure 3.32 Enantiomers of butan-2-ol. Data from *J. Am. Chem. Soc.*, 1983, **105**, 6174.

Given this property of different enantiomers, the symbols for the specific rotation (+/− or *d/l*) are often used as an additional or alternative descriptor to *R* and *S*; however, it should be noted that these descriptors tell us nothing about the spatial arrangement of atoms and so we recommend sticking to *R* and *S*. It should also be noted that there is no general correlation between *R* and *S* (or D and L discussed next) and (+/−) or d/l and that the specific rotation for a molecule is only found through empirical experimentation.

One other common stereochemical descriptor is the use of D and L (note the small capital letters, distinct from the lower-case *d* and *l*, discussed previously) prefixes. Often found in biological molecules such as sugars *e.g.* D-glucose and amino acids *e.g.* L-alanine, these prefixes originate from a naming system devised by the chemist Emil Fischer and relate to the precise stereochemical arrangement of atoms relative to the naturally occurring D-glyceraldehyde. The system is now largely obsolete beyond sugars, amino acids and a few other simple molecules.

> **▘ Checkpoint**
>
> You should now understand how we measure optical rotation of compounds.

3.4 Mixtures of Enantiomers

If a chemical reaction that has chiral products has no preferential bias for forming one enantiomer over another, statistically it will give rise to a 1:1 mixture. This mixture is known as a **racemic mixture** or a **racemate**. An example of such a reaction is the reduction of butanone with sodium borohydride to give racemic butan-2-ol, shown in Figure 3.33. If you make a model of butanone (Figure 3.33), you can see why the nucleophile can attack from both faces of the carbonyl bond with equal likelihood.

>
>
> The process of biasing the formation of one enantiomer over another is called asymmetric synthesis or enantioselective synthesis and is a major achievement in the field of organic chemistry. Read about this in organic chemistry textbooks.

> **❗ Key Learning Point**
>
> A racemic mixture or racemate is a 1:1 mixture of enantiomers with a specific rotation of zero.

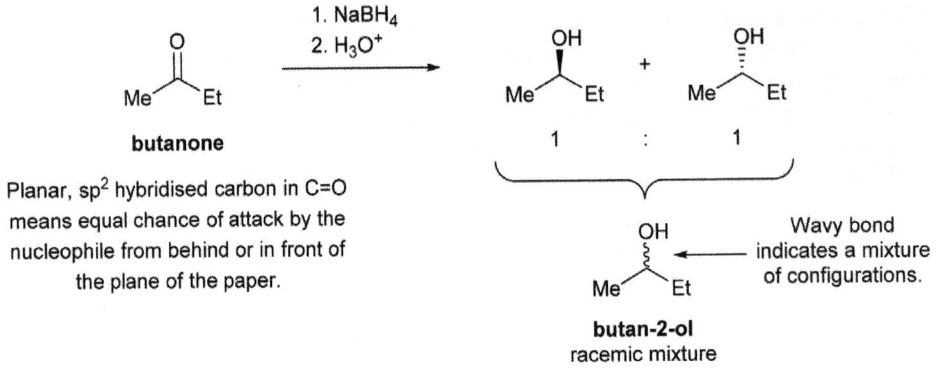

Figure 3.33   Racemic mixtures.

If we have a racemic mixture, such as that described in Figure 3.33, in addition to the wavy bond shown in the structure (denoting a mixture of configurations, R and S), we can also use stereochemical descriptors to show we have a 1:1 mixture of enantiomers. Thus, the product from the reaction can be given the name (±)-butan-2-ol or (*dl*)-butan-2-ol to show that we have a racemate. Since the two enantiomers have equal and opposite specific rotations (Figure 3.33), for a racemic mixture, we find that these cancel each other out and so $[\alpha]_D = 0$.

If we have a mixture of enantiomers that is not a racemate, we can either quote the ratio of enantiomers *e.g.* 60:40 or we can quote the amount of excess of the major enantiomer, known as enantiomeric excess or ee (again typically quoted as a percentage) calculated using the formula (for a pair of enantiomers of ratio A:B, where A is the major component)

$$\% \text{ ee} = \frac{A - B}{A + B} \times 100\%$$

This means that a mixture with an enantiomeric ratio A/B of 60:40 will have an ee of 20%.

For a non-racemic mixture of enantiomers, we get a net optical rotation with the direction of rotation consistent with the enantiomer that is in excess. We can use this to work out the ee (or optical purity here) of a mixture of enantiomers if we compare it with the optical rotation of the single pure enantiomer through the formula

$$\% \text{ ee (or optical purity here)} = \frac{[\alpha] \text{ observed}}{[\alpha] \text{ pure enantiomer}} \times 100\%$$

3.5 Biological Effects of Chirality

As we discussed previously (Figure 3.25), the biological building blocks amino acids are chiral (except glycine) and present naturally as single enantiomers. We find that other biological building blocks, such as sugars, are also chiral molecules. It follows then that the biological polymers composed of these building blocks, namely proteins, carbohydrates and nucleic acids (and also some lipids) also have an associated chirality and furthermore, since predominantly L-amino acids and D-sugars are naturally occurring,[†] the biopolymers are found to be homochiral (they only have one form of chirality). This means that for natural receptors and enzymes (largely comprised of proteins), the associated homochirality leads to different interactions with different enantiomers of client molecules. This can be thought of as similar to the interaction between your hands and a pair of gloves (Figure 3.34). Your right hand will fit well into a right-handed glove, but your left hand will not. With a pair of enantiomers, the R enantiomer will fit differently into the active site of a receptor or enzyme from the S enantiomer. Indeed, one enantiomer may not fit well at all (like the gloves!).

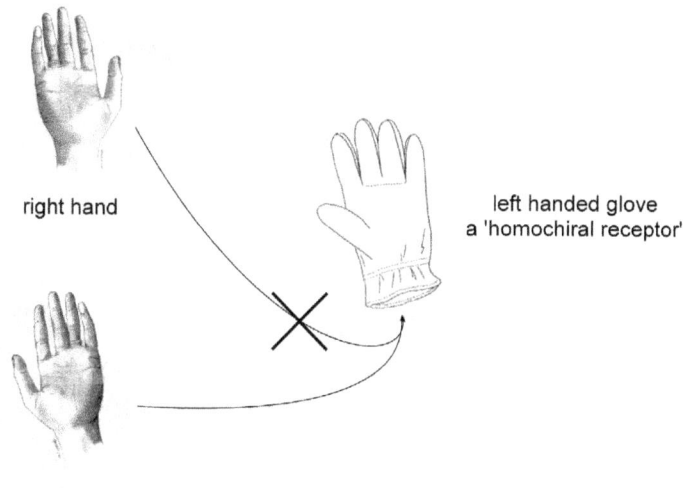

right hand

left handed glove
a 'homochiral receptor'

left hand

Figure 3.34 Homochirality in biological receptors.

This is best demonstrated with examples from fragrances and the pharmaceutical industry (Figure 3.35). The two enantiomers of limonene smell like either lemons (S enantiomer) or oranges (R enantiomer), owing to the different interactions

---

[†]There are some instances of D-amino acids being isolated from Nature.

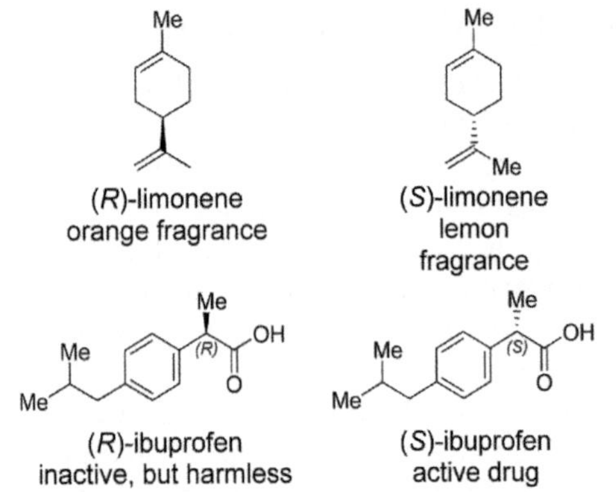

Figure 3.35 Stereoisomers of limonene and ibuprofen.

with the olfactory receptors in the nose. In the case of pharmaceuticals, only the S enantiomer of ibuprofen is found to be an active analgesic (painkiller). The R enantiomer is inactive, and is harmless in this instance, so ibuprofen can be administered safely as a racemate. Since the other enantiomer of a chiral drug will not always be benign, it is very important to understand the biological response of *both* enantiomers (particular as stereoisomers frequently have interconversion pathways in the body) to avoid unfortunate side effects, such as those associated with the morning sickness treatment thalidomide, used in the 1950s and 1960s (see Chapter 1).

3.6 Enantiomers and Diastereoisomers

So far, all the examples we have considered have contained only one stereogenic centre and so we have only had to consider two stereoisomers, enantiomers. Very frequently, molecules contain two or more stereogenic centres and this gives rise to a whole new set of potential stereoisomers. With the notable exception of stereoisomers referred to as **meso-**, discussed in Section 3.8, the number of stereoisomers for a molecule with n stereogenic centres can be calculated using the formula

$$\text{Number of stereoisomers} = 2^n$$

For a molecule with two stereogenic centres, there are four stereoisomers. Since there can only be one mirror image, only two of these stereoisomers will be enantiomers, leaving two further stereoisomers that are *not* enantiomers. These are given the name diastereoisomers (or diastereomers) and are defined as non-superimposable, non-mirror images.

> **❗ Key Learning Point**
>
> Diastereoisomers are stereoisomers that are non-superimposable, non-mirror images.

Consider the stimulant and blood pressure drug, ephedrine (Figure 3.36). The molecule contains two stereogenic centres and thus has four stereoisomers in total. Have a go at making a model of ephedrine and make sure you can assign the configuration of the two stereogenic centres correctly.

Figure 3.36 Ephedrine and its enantiomer.

Hopefully, you assigned R at the 1-position (adjacent to the alcohol group) and S at the 2-position (adjacent to the amine group). If you find this difficult, you can refer back to Section 3.2 on assigning stereo configuration.

Now make a model of the enantiomer (mirror image) of ephedrine (Figure 3.36) and superimpose the groups in the plane of the paper (the phenyl and methyl groups). This allows us to directly compare the configuration of the stereogenic centres. In the enantiomer of ephedrine, the configuration of both of the stereogenic carbons has inverted, so we now have S at the 1-position (adjacent to the alcohol group) and R at the 2-position (adjacent to the amine group). To draw the enantiomer of a compound, we can use this as an alternative approach to drawing the reflected mirror image (like in Figure 3.8): *to draw the enantiomer, invert the configuration of all stereogenic centres.* This can be as simple as making the OH group, which points to the back with a hashed bond in ephedrine, point to the front with a wedged bond and *vice versa* for the amine group. Of course, if you are in any doubt, assign the configurations as R/S to check.

Looking at Figure 3.37, let's imagine we now take ephedrine and invert the configuration of only one of the stereogenic centres, for example, the 1-position (adjacent to the alcohol group), to give the (1S, 2S)- stereochemistry, a molecule known as pseudoephedrine. If you make a model of pseudoephedrine and try to superimpose it, first on ephedrine and then on the ephedrine enantiomer that you have made previously, you find that it is neither superimposable on ephedrine nor its mirror image and so fits our definition of a diastereoisomer.

> 💬 To draw a diastereoisomer, invert the configuration of *at least one, but not all* stereogenic centres.

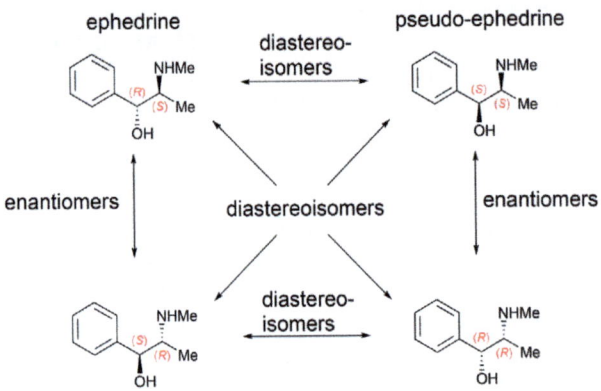

Figure 3.37 Enantiomers and diastereoisomers.

To complete the set of ephedrine stereoisomers, let's make a model with the (1*R*, 2*R*) stereochemistry. Again, if we attempt to super-impose it on ephedrine itself or its enantiomer, we find neither is achievable and so we have another diastereoisomer. If we consider this new molecule's relationship with pseudoephedrine, we find that it is the mirror image and hence is its enantiomer (check that you are happy that the configuration of *both* stereogenic centres of pseudoephedrine have been inverted). The relationship between all four of the stereoisomers of ephedrine is shown in Figure 3.37. Overall, we can say that each stereoisomer of ephedrine has one enantiomer but two diastereoisomers.

Let's check you've understood diastereoisomerism by drawing out all the stereoisomers of the molecule in Figure 3.38, indicating the stereochemical relationship (*i.e.* enantiomer or diastereoisomer).

To draw the enantiomer, invert *all* stereogenic centres.

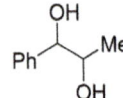

Figure 3.38 Can you identify pairs of enantiomers and diastereoisomers?

Hopefully you drew out four different structures related to each other as shown in Figure 3.39.

3.7 Properties of Diastereoisomers

For a pair of enantiomers, such as (*R*)- and (*S*)-butan-2-ol, we found that they had identical physical and spectroscopic proper-ties, apart from their optical rotation (Figure 3.32). If we consider a pair of diastereoisomers, we now find that they have different

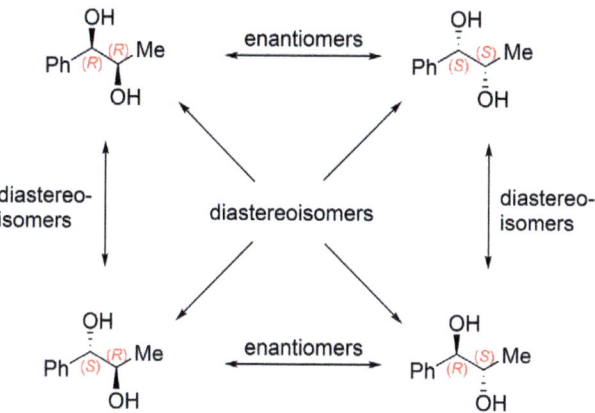

Figure 3.39 Two pairs of enantiomers and four pairs of diastereoisomers.

physical, spectroscopic and chemical properties and are often easily separated through processes such as distillation, crystallisation or chromatography. If we consider the properties of the four stereoisomers of ephedrine (Figure 3.40), we find that ephedrine and its enantiomer have, as expected, identical melting points, but equal yet opposite specific rotations. If we now consider ephedrine *vs.* pseudoephedrine, we find completely different melting points (40 °C and 118–118.5 °C, respectively) and different optical rotations. The two diastereoisomers also exhibit different NMR spectra, exemplified by the proton at the 1-position (adjacent to the alcohol group, indicated in red in Figure 3.40), which has a chemical shift δ of around 4.9 ppm for (−)-ephedrine and around 4.3 ppm for (+)-pseudoephedrine (data from *Chem. Pharm. Bull.*, 2005, **53**, 105).

> **◼ Checkpoint**
>
> You should now understand the differences in the properties of enantiomers and diastereoisomers.

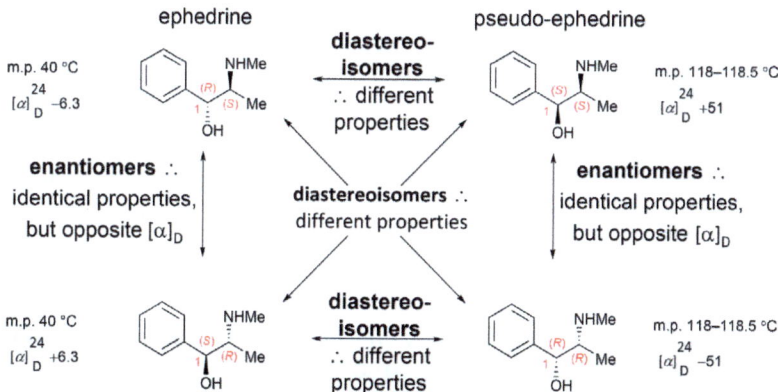

Figure 3.40 Properties of enantiomers and diastereoisomers. Melting point data from *Pol. J. Chem.*, 1985, **59**, 395 and *Justus Liebigs Ann. Chem.*, 1929, **470**, 174. Optical rotation data from *Heterocycles*, 2001, **55**, 1455.

3.8 *meso* Compounds

> **❗ Key Learning Point**
>
> A *meso compound* is one containing two or more stereogenic atoms but that is superimposable on its mirror image and therefore is achiral.

Previously, we discussed the general formula 2^n for the number of stereoisomers for a compound containing n stereogenic centres, but indicated that there was one exception to the rule: *meso* compounds. A *meso* compound contains two or more stereogenic atoms but is superimposable on its mirror image and so is achiral, with a specific rotation $[\alpha]_D = 0$.

Please make models of the natural products (2*S*,3*S*)-(+)-tartaric acid (commonly found in grapes) and its enantiomer, (2*R*,3*R*)-(−)-tartaric acid, shown in Figure 3.41. As with any other pair of enantiomers, you find that you cannot superimpose the two compounds and they each have an associated specific rotation; −13.5 and +13.5, respectively.

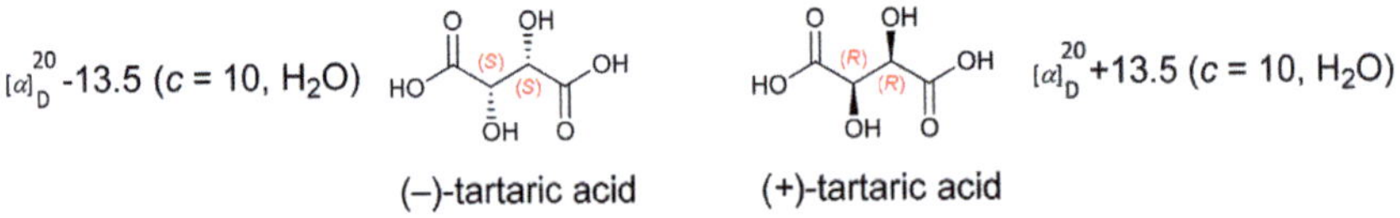

$[\alpha]_D^{20}$ -13.5 (c = 10, H$_2$O)

(−)-tartaric acid   (+)-tartaric acid

$[\alpha]_D^{20}$ +13.5 (c = 10, H$_2$O)

Figure 3.41 Chiral enantiomers of tartaric acid. Data from *J. Fluor. Chem.*, 1980, **15**, 191.

💬 An easy way to spot the possible existence of a mirror plane is if you notice that the substituent groups on each stereogenic centre are the same, e.g. H, OH and CO$_2$H in the tartaric acid example.

Now make models of (2*R*,3*S*)-tartaric acid and (2*S*,3*R*)-tartaric acid (Figure 3.42), both diastereoisomers of (2*R*,3*R*)-(−)-tartaric acid. Experimentally, we observe a specific rotation of 0°, suggesting something unexpected is occurring. If we attempt to superimpose the two molecules, we find that this is now achievable and so, although the compounds look at first sight as though they should be enantiomers of one another, they are in fact the same stereoisomer and so this is a *meso* compound.

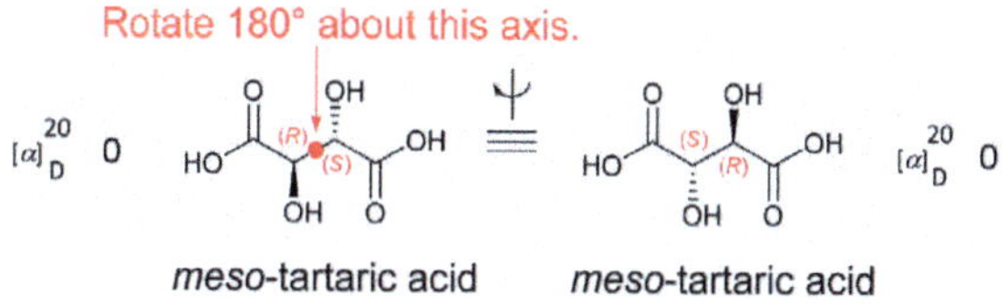

Rotate 180° about this axis.

meso-tartaric acid   *meso*-tartaric acid

Figure 3.42   *meso* stereoisomer of tartaric acid. Imagine sticking a pin through the paper at the red dot and spinning the molecule around that axis.

We need not make two models in order to spot *meso*-compounds, because they contain an internal mirror plane, as shown in Figure 3.43. Rotate the central C–C single bond of the (2*R*,3*S*)-tartaric acid model so that you have the substituents in the positions shown (also known as conformational isomers – see Chapter 4). Now you can clearly see a plane of symmetry (or mirror plane) bisecting the central C–C bond, indicating that we have a *meso*-stereoisomer.

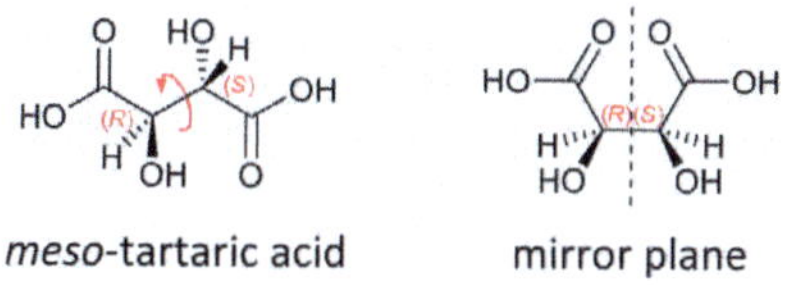

meso-tartaric acid   mirror plane

Figure 3.43   Spotting *meso*-compounds through the internal mirror plane.

■ **Checkpoint**

You should now understand the definition of pairs of enantiomers, pairs of diastereoisomers, racemates (racemic mixtures) and *meso*-compounds.

Overall, the relationship of the three stereoisomers of tartaric acid is shown in Figure 3.44.

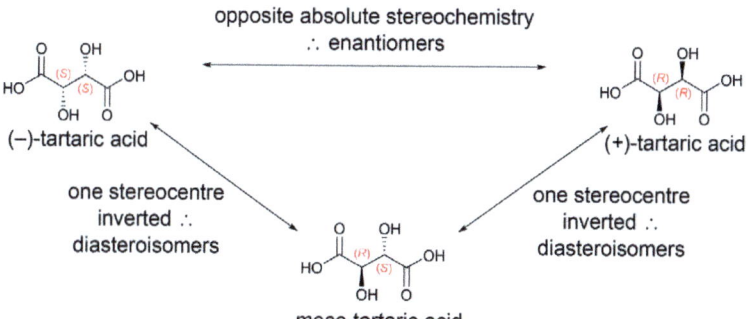

opposite absolute stereochemistry
∴ enantiomers

(−)-tartaric acid   (+)-tartaric acid

one stereocentre inverted ∴ diastereoisomers

one stereocentre inverted ∴ diastereoisomers

meso-tartaric acid

Figure 3.44   Stereoisomers of tartaric acid.

Let's make sure we've understood what *meso*-isomers are by drawing out all the stereoisomers of the molecule shown in Figure 3.45, indicating the stereochemical relationship (*i.e.* enantiomer, diastereoisomer or *meso*-).

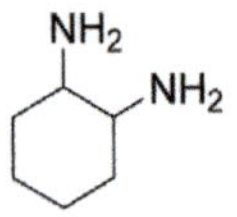

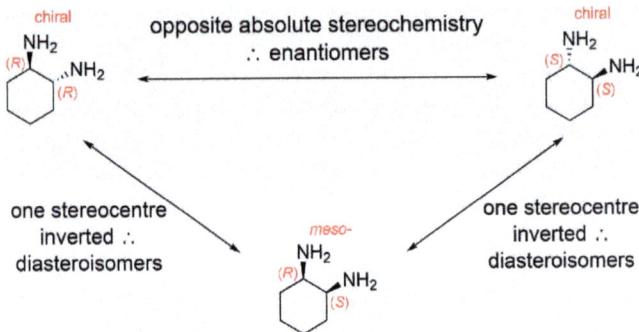

Figure 3.45 Can you find all the stereoisomers of 1,2-diaminocyclohexane?

The answer is given in Figure 3.46.

Figure 3.46 Two chiral enantiomers and a *meso*-diastereoisomer.

3.9 Pseudochirality

Let's now consider the molecule ribaric acid, shown in Figure 3.47. The carbon marked with an asterisk is, strictly speaking, not a stereogenic centre because the connectivity of atoms in the two groups circled in red is the same. However, if the configuration of the two actual stereogenic carbons is considered (make sure you can assign these correctly using the CIP rules as discussed previously), we find that one carbon has the R configuration and the other has the S configuration. This means that, although not constitutional, there is some difference between the two substituents around the

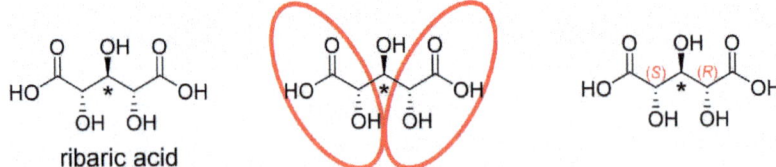

ribaric acid

Figure 3.47 Pseudochirality.

asterisked carbon. This phenomenon is known as **pseudochirality**. We can assign stereochemical descriptors for pseudochirality in the same way as for normal stereogenic atoms, but rather than *R/S* stereodescriptors denoting clockwise or anticlockwise movement from priorities $1 \rightarrow 2 \rightarrow 3$, respectively, we use lower-case *r/s* (for clockwise and anticlockwise, respectively) to denote that we have pseudochirality.

Make a model of ribaric acid as depicted in Figure 3.48. Using the CIP rules, from Step 1, we find that O has the highest atomic number (8) so takes priority 1 and H has the lowest atomic number (1) so takes priority 4. Now we must consider our two groups that are identical in terms of connectivity but different in terms of configuration. For pseudochirality, *R* configuration takes priority over *S* configuration, so the *R*-configured carbon atom takes priority 2 and the *S*-configured carbon atom takes priority 3. We can now carry on as before; rotating the molecule to look down the C–H (priority 4) bond (Step 2) and looking at priority $1 \rightarrow 2 \rightarrow 3$ (Step 3), we find we are moving in a clockwise direction and so we have the *r*-pseudostereochemistry, (2*R*,3*r*,4*S*)-ribaric acid.

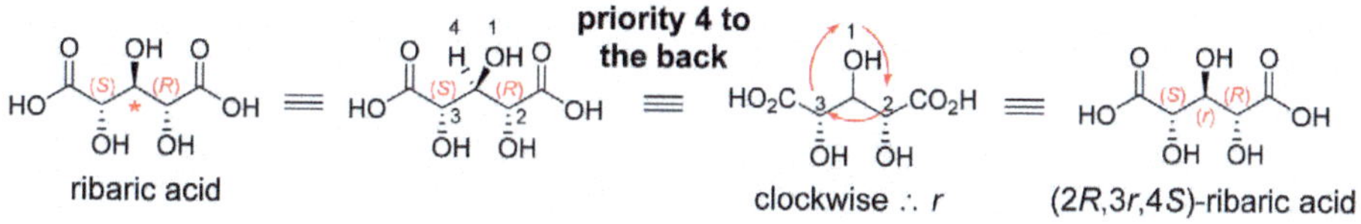

Figure 3.48 Pseudochirality stereochemical descriptors.

3.10 Conclusion

Well done! You should now understand what makes molecules chiral or achiral and be able to assign the configuration of a stereocentre as *R* or *S*. You should be really comfortable in spotting pairs of enantiomers, pairs of diastereoisomers, racemates (racemic mixtures) and *meso*-compounds.

In the next chapter you will explore the concept of conformation (rotation around single bonds) and its effect on energy and NMR spectroscopy, while learning about an important new way to depict molecules in 3D: Newman projections. Do come back to this chapter to remind yourself of the key learning points as you proceed.

Exercises

The first four questions are mainly to check your understanding, though they could be found as parts of exam questions. A sample exam question then follows.

1. Assign stereochemical descriptors for the stereogenic centres in the following molecules.

(i)   (ii) 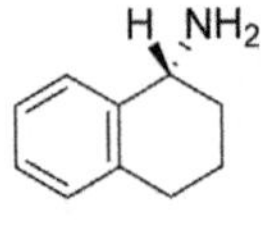   (iii) 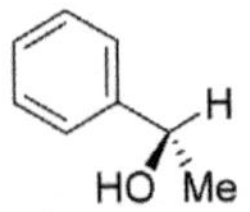   (iv)

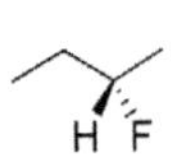

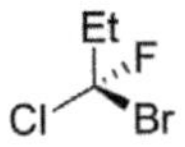

(v) 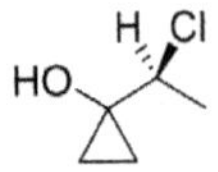   (vi) 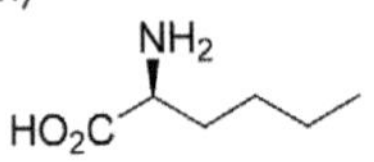   (vii) 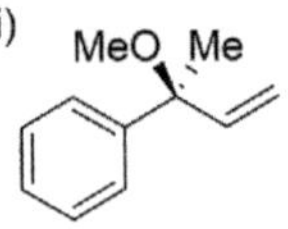   (viii)

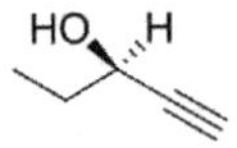

(ix) 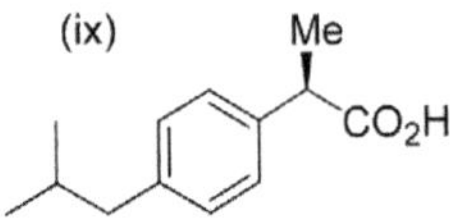   (x) 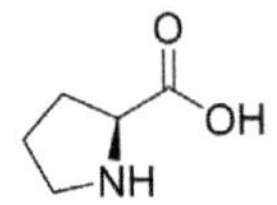   (xi) 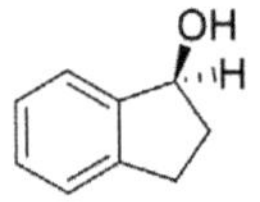   (xii)

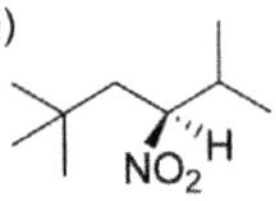

(xiii) 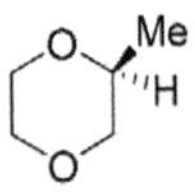   (xiv) 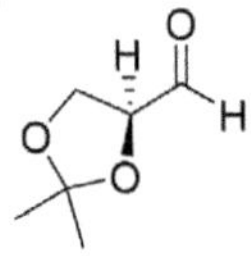   (xv) 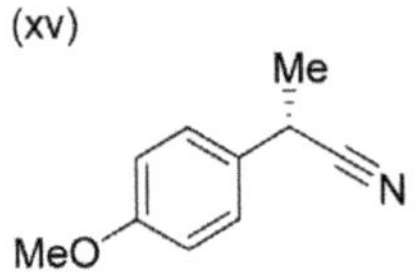   (xvi)

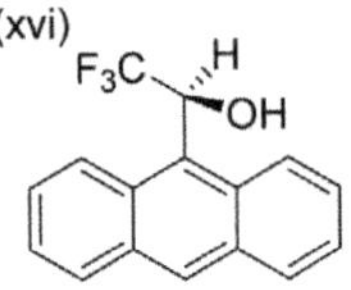

2. Compare the following compounds and decide if they are an enantiomer, diastereoisomer, or the same as the diol structure in the box.

> 💬 You may find it helpful to work out the absolute stereochemistry at each stereocentre.

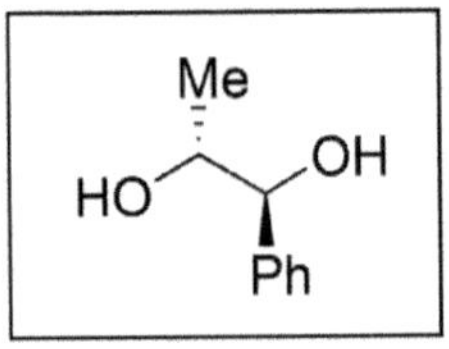

(i) 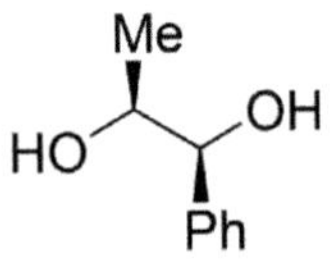   (ii) 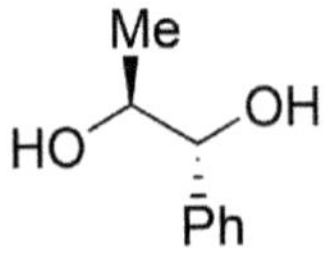   (iii) 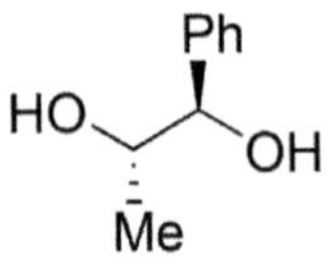   (iv)

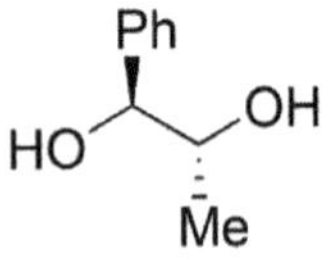

3. Assign the configuration and pseudochirality to the following molecules.

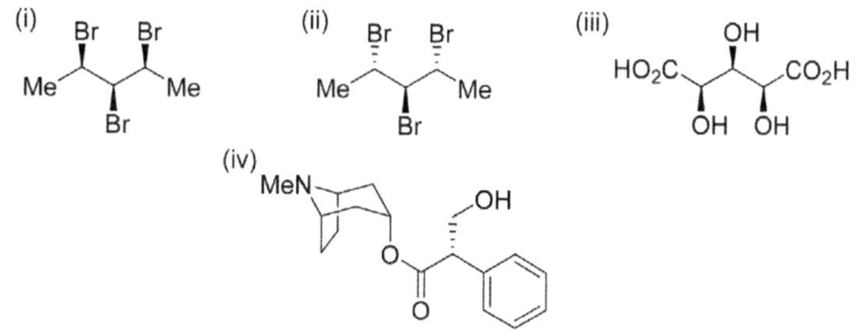

4. Draw out all the stereoisomers of the following molecules, indicating the stereochemical relationship (*i.e.* enantiomer, diastereoisomer, *meso*-).

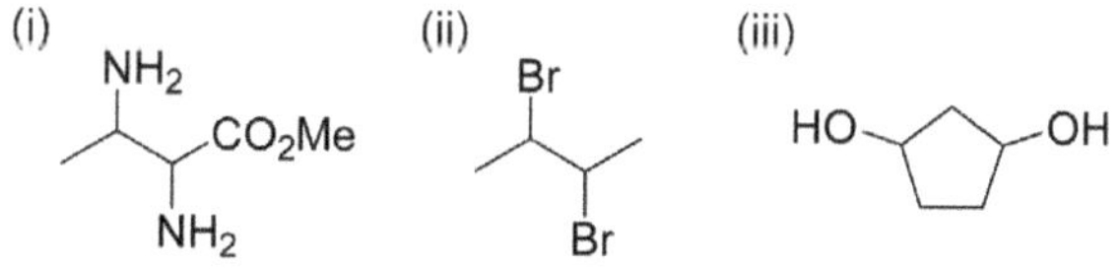

? **Example Exam Question**

This example question is chosen to test your understanding of the concepts covered in this chapter. It covers all areas discussed in the chapter, as well as related concepts from previous chapters.

Consider the stereoisomer of pentane-2,4-diol shown in Figure 3.49 and answer the following questions:

i. Use the Cahn–Ingold–Prelog rules to define the absolute configuration of the two stereogenic centres.
ii. Draw the enantiomer of the stereoisomer shown in Figure 3.49.
iii. Draw a diastereoisomer of the molecule shown in Figure 3.49.
iv. Indicate which of the stereoisomers you have drawn are chiral molecules and which are meso-compounds, justifying your answer with suitable diagrams.
v. Indicate which of the stereoisomers will exhibit a specific rotation, ($[\alpha]_D \neq 0$).

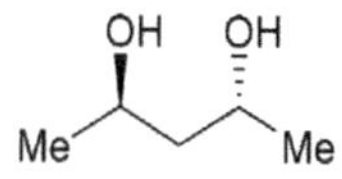

Figure 3.49   Pentane-2,4-diol.

Answers to Exercises

1.

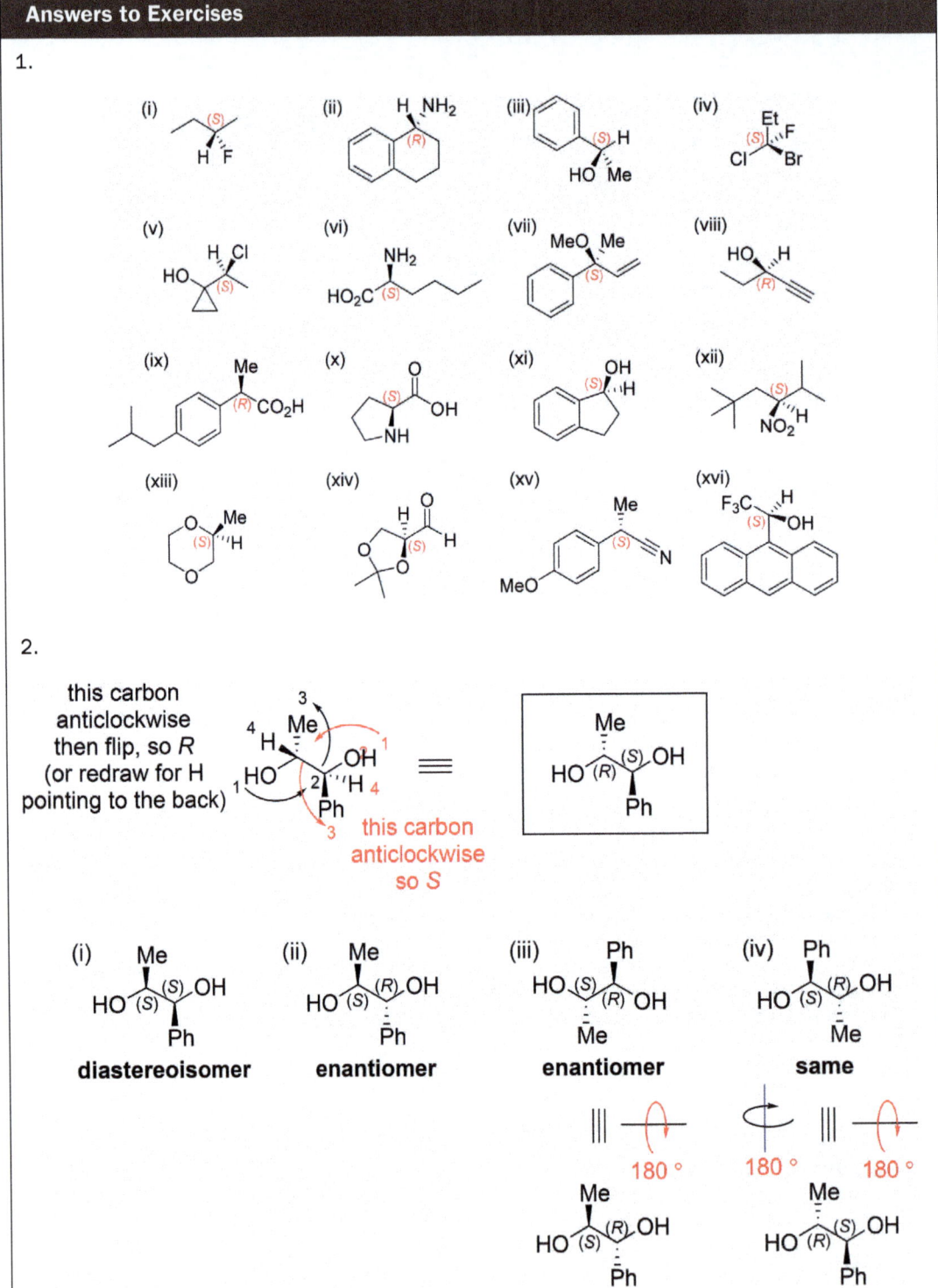

2.

this carbon
anticlockwise
then flip, so *R*
(or redraw for H
pointing to the back)

this carbon
anticlockwise
so *S*

3.

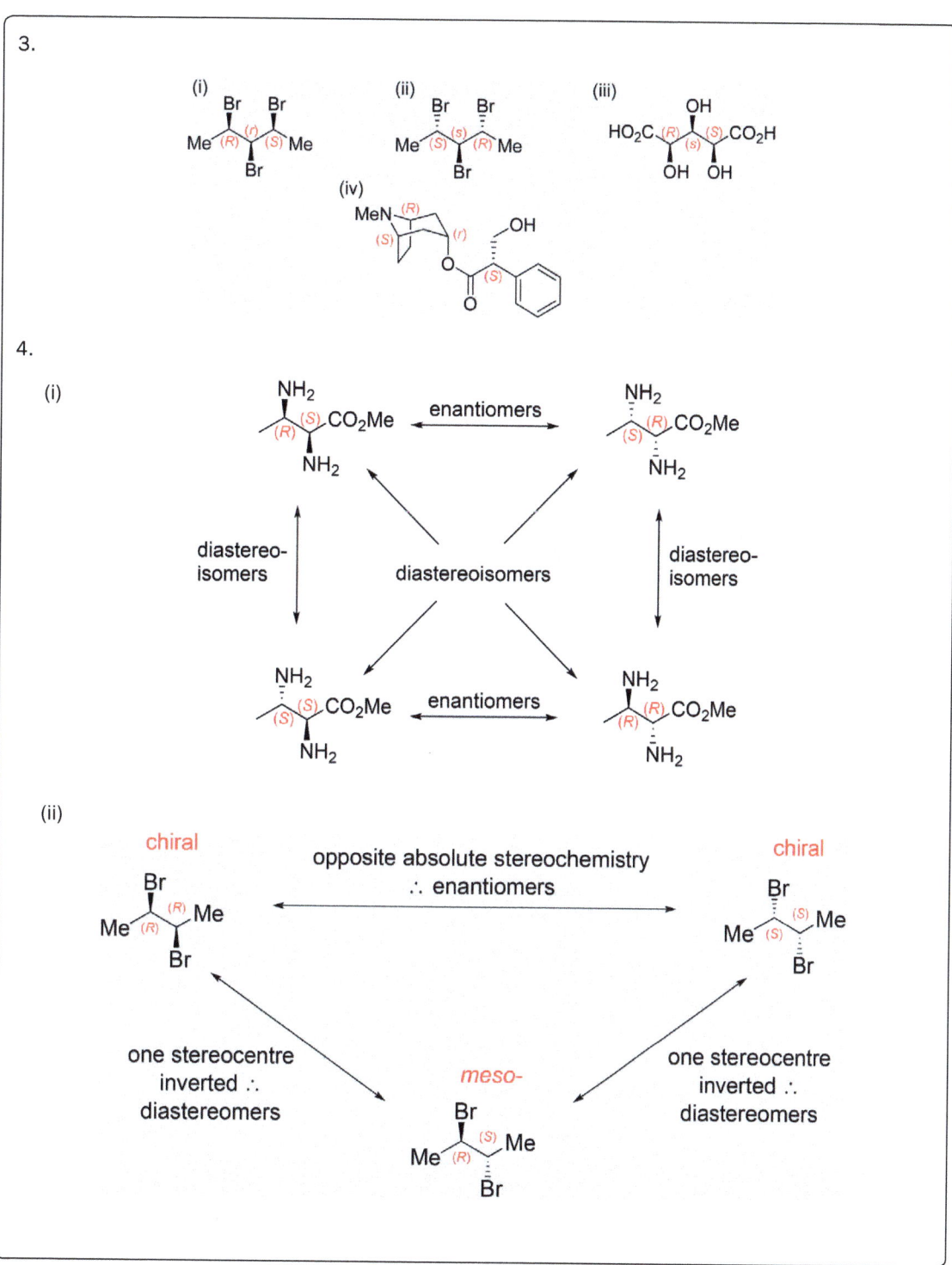

(i)

Br   Br
Me (R) (r) (S) Me
         Br

(ii)

Br   Br
Me (S) (s) (R) Me
         Br

(iii)

          OH
HO₂C (R)   (S) CO₂H
     (s)
      OH   OH

(iv)

MeN (R)
   (S)   OH
        O   (r)
        O   (S)

4.

(i)

NH₂
(R) (S) CO₂Me
   NH₂

⟷ enantiomers ⟷

NH₂
(S) (R) CO₂Me
   NH₂

diastereo-isomers

diastereoisomers

diastereo-isomers

NH₂
(S) (S) CO₂Me
   NH₂

⟷ enantiomers ⟷

NH₂
(R) (R) CO₂Me
   NH₂

(ii)

chiral

Br
Me (R) (R) Me
         Br

⟷ opposite absolute stereochemistry ∴ enantiomers ⟷

chiral

Br
Me (S) (S) Me
         Br

one stereocentre inverted ∴ diastereomers

meso-

Br
Me (R) (S) Me
         Br

one stereocentre inverted ∴ diastereomers

(iii)

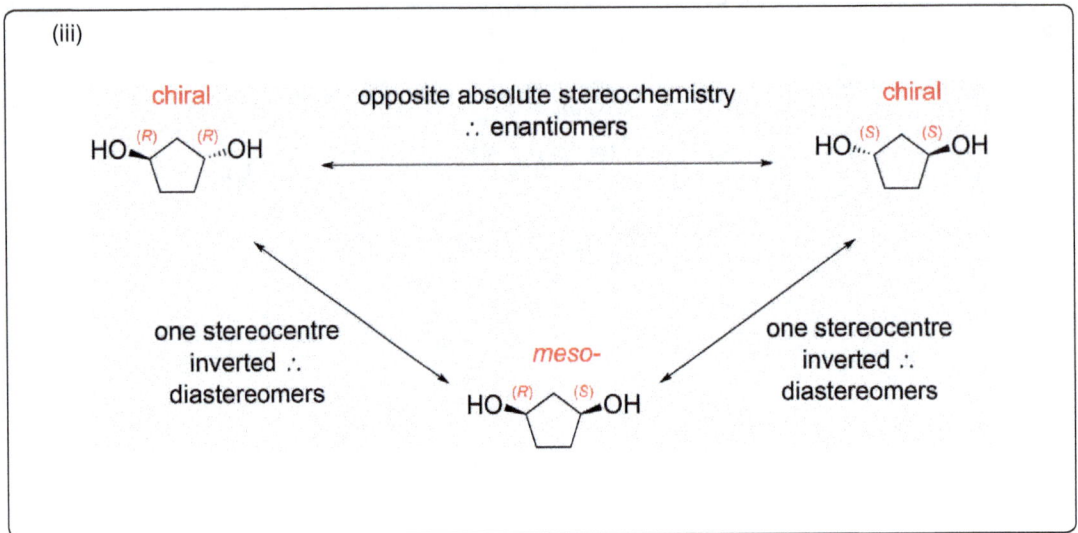

? Answer to Example Exam Question

i.

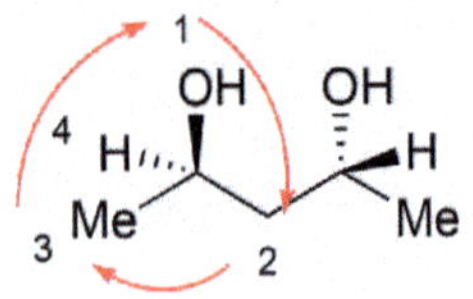

clockwise ∴
R configuration

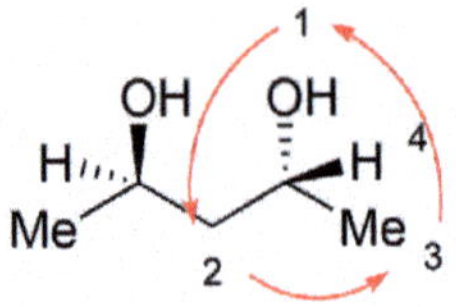

Priority 4 is to the front.
Proceed as normal; anticlockwise
gives *S*, but reverse to compensate
∴ *R* configuration.

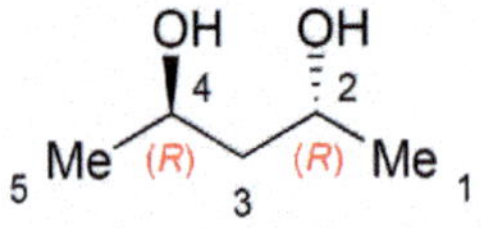

(2*R*,4*R*)-pentane-2,4-diol

ii. Invert all stereocentres to make the enantiomer:

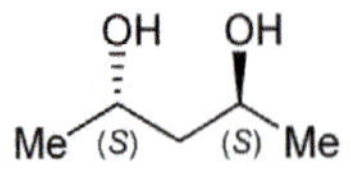

(2*S*,4*S*)-pentane-2,4-diol

iii. Invert at least one, but not all, stereocentres to make a diastereoisomer, so one of two stereocentres in this case:

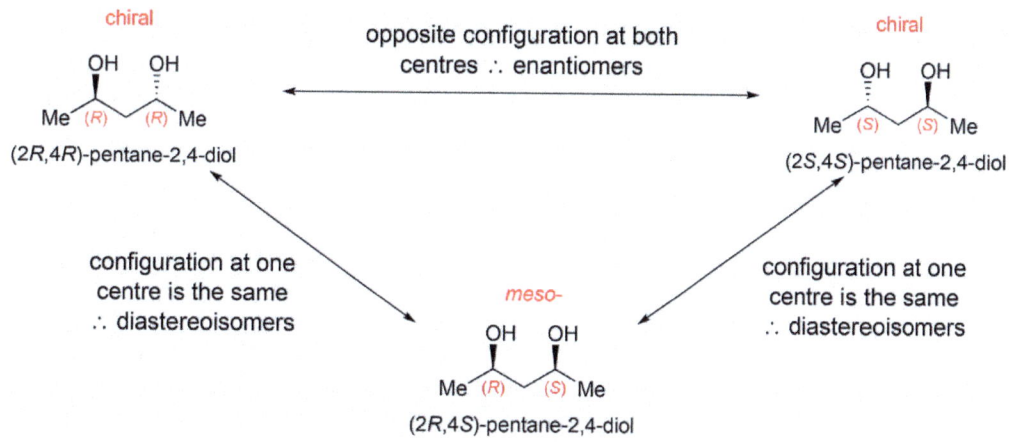

iv.

(2R,4S)-pentane-2,4-diol has a mirror plane and is superimposable on its mirror image, so is *meso*-.

v. The chiral stereoisomers will be optically active, so (2R,4R)-pentane-2,4-diol and (2S,4S)-pentane-2,4-diol will exhibit a specific rotation, while the *meso*-stereoisomer will not.

By the End of This Chapter You Will:

- Understand the difference between configuration and conformation.
- Be able to draw Newman projections and translate between them, 2D depictions and sawhorse projections.
- Understand what factors stabilise or destabilise acyclic conformations.
- Understand the meaning of *anti* and *syn* relative stereochemistry.
- Understand that conformation influences coupling constants in ^{1}H NMR spectroscopy.
- Be able to predict the coupling constant by considering the dihedral angle between two proton nuclei.
- Understand what is meant by the terms enantiotopic and diastereotopic and how these properties affect ^{1}H NMR spectra.

What You Will Get from This Chapter

At first sight, introduction of 'sawhorse projections' and 'Newman projections' may seem like we're making everything more complicated. But no, these representations are your friends! They make it easier to explain 3D chemistry on a 2D page – we're just looking at the same molecule from various viewpoints. So the secret to success in Chapter 4 is to build your confidence in drawing molecules in these different ways through practice and more practice. And yes, of course, your molecular model kit will make it all much easier to see.

Conformation of Acyclic Compounds

This chapter builds on that covered previously and introduces new ways to depict a molecule by looking down a specific bond (Newman projections, briefly mentioned in Chapter 1, and saw-horse projections), starting with acyclic examples and leading to cyclic examples in Chapter 5. This is particularly useful for considering molecules in 3D and the implication of rotating around bonds to give different conformational isomers, also known as conformers (conformational isomers are those where one or more single bonds have been rotated). It will be very helpful as you become more familiar with manipulating molecules in your head, a particularly key skill that is often tested in intermediate- and higher-level exams at University. When you start out this may be a difficult concept to grasp, so we recommend using a model kit as an aid, until you get more proficient in mental rotation in 3D.

---

❶ Key Learning Point

Conformers (or conformational isomers) are species that can be interconverted through rotation of one or more single bonds.

4.1 Newman Projections

Let's consider the standard 2D depiction that we have been using in the book so far for the molecule (3S,4S)-4-amino-2,2-dimethylpentan-3-ol (Figure 4.1). Introduced briefly in Chapter 1, a Newman projection is merely a different way of depicting the same molecule by concentrating on a view down a particular bond from a given direction. The depiction is very useful for considering relative energies for the different conformations a molecule can adopt, reactivities of particular conformations, stereochemical outcomes *e.g.* for elimination reactions and also the magnitudes of coupling constants (J) in NMR spectroscopy. This is because the angle (known as the dihedral or torsion angle, ϕ), across three bonds or between two adjacent

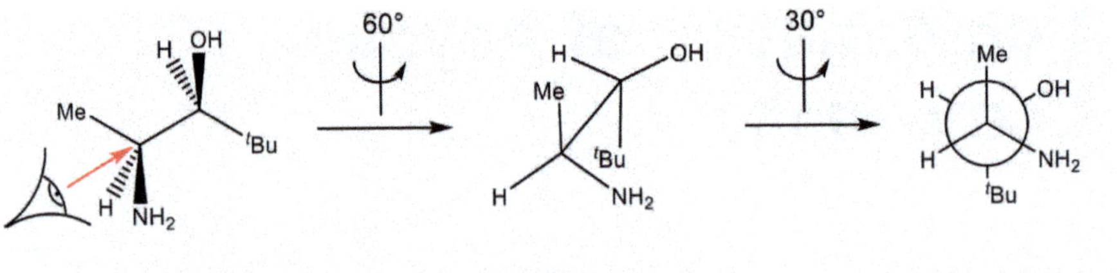

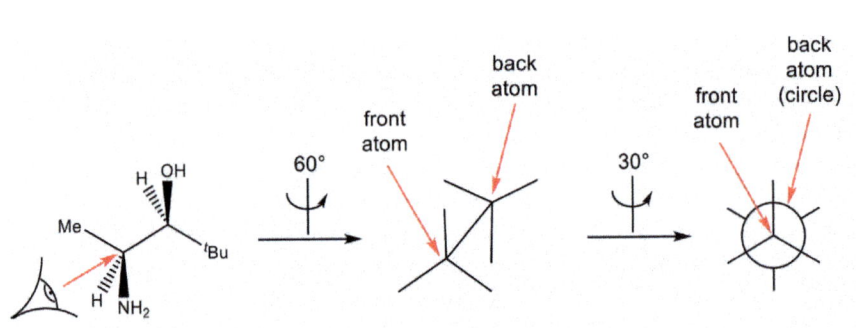

2D conformation   sawhorse projection   Newman projection

Figure 4.1   Sawhorse and Newman projections.

substituents is most easily visualised when looking directly down the central bond. A Newman projection is depicted where the central atom closest to you (in the given view) is represented by the junction where the four bonds meet, and the furthest atom is represented by a circle with the three other bonds off that circle.

To aid in the correct translation from the 2D depiction to the Newman projection, we will include an intermediate sawhorse projection in which the molecule has been rotated by ≈60° about the z-axis from the initial 2D depiction, whereas the Newman projection has been rotated by 90° (Figure 4.1). Once you are comfortable with converting 2D depictions to Newman projections, there is no need to go *via* the sawhorse projection.

4.2   How to Draw Newman Projections

A good bit of advice for drawing sawhorse and Newman projections (and chair conformations in Chapter 5) is, in the words of Dr David Fox, '*Don't decorate the Christmas tree before you put it up.*' Make a model of (3*S*,4*S*)-4-amino-2,2-dimethylpentan-3-ol, ensuring the stereochemistry at the two stereogenic carbons is consistent with that depicted in the 2D representation. Now draw blank sawhorse and Newman projection carbon skeletons, as shown in

> Often once you've generated your Newman projection, you will need to change it to another conformation. It may be tempting to do this while drawing your sawhorse and Newman projections, but we suggest not rushing it.

2D conformation   sawhorse projection   Newman projection

Figure 4.2   Blank sawhorse and Newman projections.

Figure 4.2. This is the 'Christmas tree' that can now be decorated with the substituents. Please note that whether the front carbon has the vertical bond pointing up or down depends on both the specific conformation of the molecule and the direction the bond is viewed down (specified by the eye in Figure 4.2). Here the front carbon has the central vertical bond pointing up and the back carbon has the central vertical bond pointing down, in a so-called **staggered** conformation.

The next step is to fill in the substituents on the bonds in the plane of the paper in the 2D conformation (bonds shown in bold red in Figure 4.3). These will translate to the vertical bond substituents in the sawhorse and Newman projections. Let's consider the front carbon first. The substituent in the plane of the paper is the methyl group, so we can fill that in as the vertical bond substituent pointing up in the sawhorse and Newman projections. Next we take the back carbon, where we can see that the substituent in the plane of the paper is the *t*-Bu group. We can fill that in as the vertical bond substituent pointing down in the sawhorse and Newman projections.

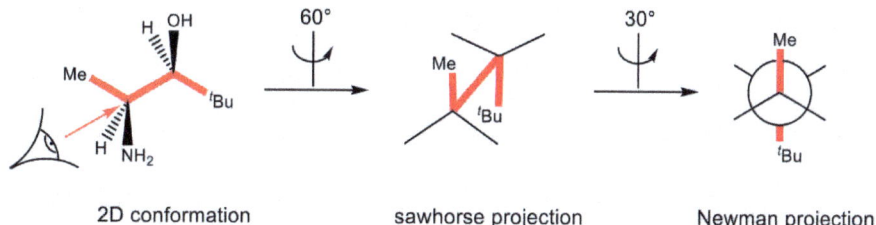

2D conformation   sawhorse projection   Newman projection

Figure 4.3   Adding the substituents in the plane of the paper to sawhorse and Newman projections.

Now comes the more error-prone process of translating the substituents away from (hashed) and towards (wedged) the viewer, to the sawhorse and Newman projections. We'll start with the sawhorse depiction. Just take your molecular model in the orientation of the 2D paper depiction and rotate it anticlockwise by 60° about the vertical axis (to get close to the viewing angle depicted by the eye). Looking at the front carbon first, we can see that the wedged bond substituent is now on the right-hand side and the hashed bond substituent is now on the left-hand side. Let's fill in these substituents in our sawhorse depiction. Now looking at the back carbon, we see that the wedged bond substituent is now on the right-hand side and the hashed bond substituent is now on the left-hand side. If we fill in these substituents in our sawhorse depiction, the finished sawhorse projection should look like that shown in Figure 4.4.

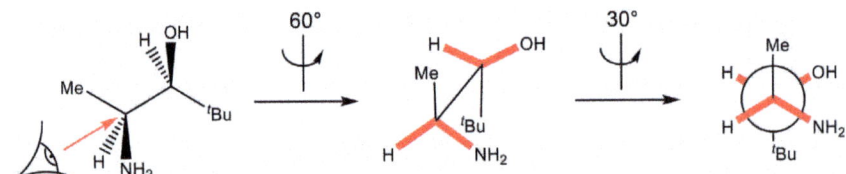

2D conformation   sawhorse projection   Newman
                                                        projection

Figure 4.4   Adding the substituents in and out of the plane of the paper to the sawhorse projection.

In an exam situation where you do not have a model kit available, use anything to hand to help get a perspective, e.g. a pen.

Let's turn our attention to the Newman projection, obtained by rotating our molecular model by a further 30° about the z-axis to get the viewing angle down the C–C single bond depicted by the eye. Once again, looking at the front carbon first, you will see that the wedged bond substituent is now on the right-hand side and the hashed bond substituent is now on the left-hand side. If we fill in these substituents in our Newman projection and look at the back carbon, we see that the wedged bond substituent is now on the right-hand side and the hashed bond substituent is now on the left-hand side. If we fill in these substituents in our Newman projection, our finished Newman projection should look like that shown in Figure 4.5.

2D conformation   sawhorse projection   Newman projection

Figure 4.5   Adding the substituents on hashed and wedged bonds to the Newman projection.

If using a pen, draw the 'undecorated' Newman projection. Hold the pen side-on; this is the 2D 'molecule' depiction. Imagine that the hashed substituents are at the back and the wedged substituents are at the front. Keeping track of where these substituents are, now rotate the pen in the direction required to look down the bond for the Newman projection.

To translate from a Newman projection to the side-on 2D representation, we use the same stepwise approach of drawing an 'undecorated' 2D framework (remembering that the bonds pointing vertically up and down in the Newman projection should correspond to those pointing up and down in the plane of the paper in the 2D), rotating the Newman projection accordingly and tracking the position of the substituents on the left and the right-hand sides of the initial projection to decide what is wedged and hashed in the 2D depiction.

Of course, a molecular model will be very helpful again!

Let's check you can do this. For the 2D molecules in Figure 4.6, in the conformer depicted, convert them to a Newman projection from the perspective of the eye shown.

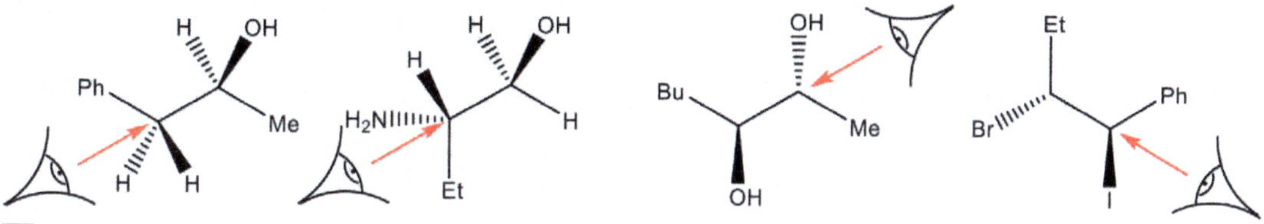

Figure 4.6 Checkpoint examples – use sawhorse intermediates to help you if you need to.

For the Newman projections shown in Figure 4.7, convert the molecules to the corresponding 2D depiction, putting the front carbon on the *left*-hand side.

Figure 4.7 Checkpoint examples – use sawhorse intermediates to help you if you need to.

Check your answers against Figure 4.8.

4.3 Considering Conformation Using Newman Projections

Earlier in the chapter, you met the term conformer or conformational isomer, which describes the relationship between isomers that are interconverted *via* the rotation of one or more single bonds. Newman projections, which you have just learnt to draw, are very useful in depicting and considering different conformers and their relative energies and, hence, stabilities. As previously mentioned, this is because the dihedral or torsion angle, ϕ, across three bonds or between two adjacent substituents is most easily visualised when looking directly down the central bond. You can imagine the angle ϕ between two substituents, A and D, being like the angle between two pages of a book, as shown in Figure 4.9. The central bond between substituents B and C is placed down the divide between the two pages; imagine that the bonds A–B and C–D are stuck to the left- and right-hand pages, respectively. Opening and shutting the book varies the dihedral angle between 0° and 180°. Of course, the limitation of this model is that a book has a limit to how far it can open, whereas a bond (at least in theory) has free rotation about 360°.

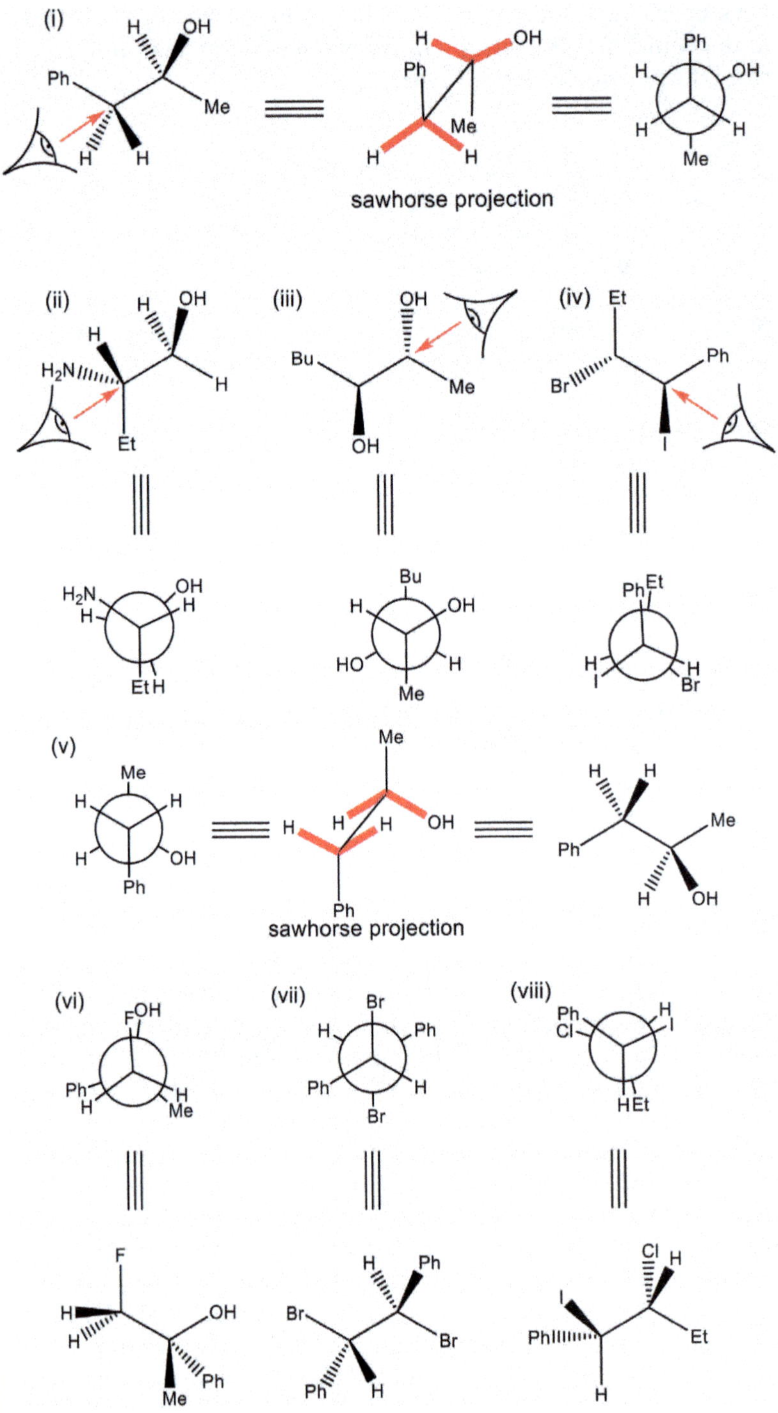

sawhorse projection

sawhorse projection

Figure 4.8 Checkpoint answers.

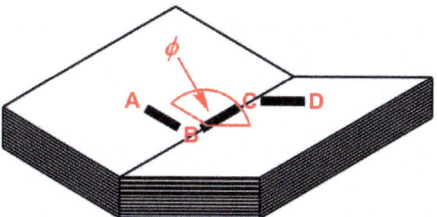

Figure 4.9   Dihedral angle.

4.4 Energies of Different Conformations

As you might imagine, not all possible conformations of a multi-bond molecule are equal; in fact, they can have significantly different energies, stabilities and reactivities, as you will see over the next two chapters. To study this, make a model of the simple molecule ethane (C_2H_6) in the conformation shown in Figure 4.10.

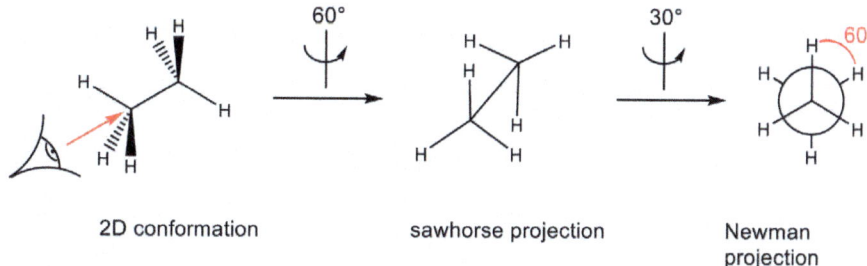

Figure 4.10   Depictions of the staggered conformation of ethane.

The conformation you have just made is known as a staggered conformation, which, when viewed from the side in 2D, is the classic **zig-zag** arrangement of the atoms in the plane of the paper. This conformation is, in fact, stabilised by both steric and orbital factors and tends to be low in energy relative to other possible conformations, which is why it is the most commonly drawn conformation for organic molecules.

Considering sterics first, if you look at your model, the hydrogens are all as far away as possible from each other (60°) and so steric repulsion is minimised.

Now considering orbitals, the key rule is that, for adjacent atoms, parallel orbitals interact the most (although this can work in favour or against the preferred energetics of the system). For the bonds in the plane of the paper, the σ orbital of the left-hand C–H bond is in parallel with the σ* orbital of the right-hand C–H bond and so can overlap and donate electron density through hyperconjugation (depicted in Figure 4.11). This lowers the energy of the electrons in the system and leads to an overall increase in the net bonding.

Almost any finding in chemical structure and reactivity can be explained by considering one or more of the following factors:

- Sterics (size)
- Electronics (charge or partial charge)
- Orbitals (size, shape and orientation).

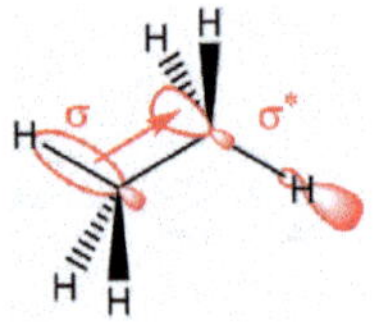

Figure 4.11   Hyperconjugation interaction stabilising the staggered conformation.

This can also occur the opposite way round (*i.e.* the σ orbital of the right-hand C–H bond can hyperconjugate with the σ* orbital of the left-hand C–H bond) and so there is significant stabilisation through hyperconjugation through the bonds in the plane of the paper. If you look at your model again, there are, of course, two further sets of parallel bonds, shown in bold black in Figure 4.12. This means that, overall, there are six hyperconjugation pathways in the ethane molecule, hence the relatively low energy of the staggered conformation.

2D conformation   sawhorse projection   Newman projection

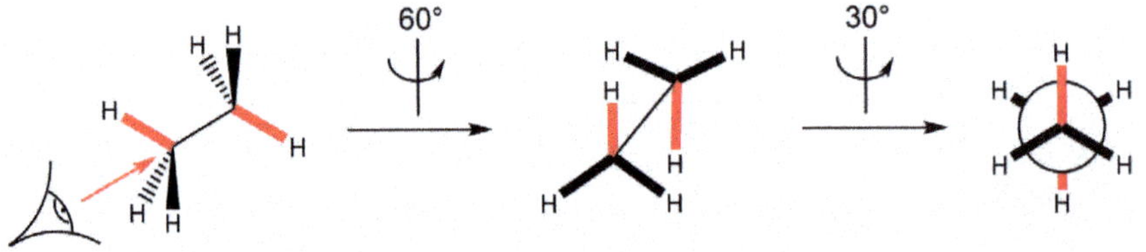

Figure 4.12   Multiple hyperconjugation pathways in ethane.

With our model of ethane, keeping the front CH_3 group in the same orientation, let's rotate the back CH_3 group through 60° so that the C–H bonds are all in parallel, as shown in Figure 4.13. This conformation is known as the **eclipsed** conformation because the bonds are all eclipsing one another.

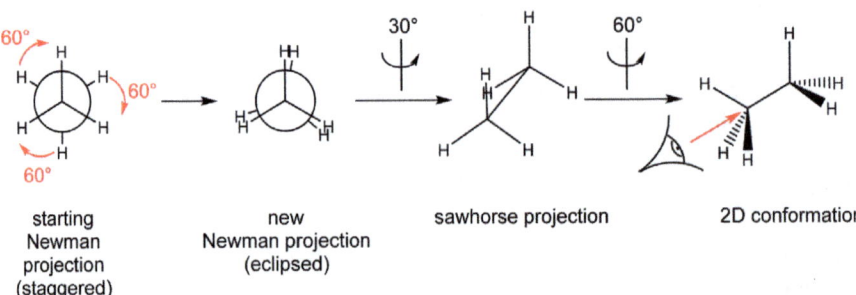

starting Newman projection (staggered)   new Newman projection (eclipsed)   sawhorse projection   2D conformation

Figure 4.13   Rotating the groups to an eclipsed conformation working from Newman to 2D depiction.

> **❗ Key Learning Point**
>
> **Torsional strain** is the increase in energy resulting from the unfavourable interaction of electrons in two separate bonds.

This conformation is the highest-energy conformation for ethane, owing to a combination of steric (minor) and electronic (major) factors.

Considering sterics first, if you look at your model, the hydrogens are now as close together as they can be (dihedral angle = 0°) and so, even though hydrogen is relatively small, there is more of a steric interaction than for other ethane conformations, so this raises the energy of the conformation (Note that this steric component is much more pronounced for groups larger than H, see later, in Figure 4.16).

Considering a combination of electronics and orbitals, for the bonds in the plane of the paper, the filled σ orbitals of the left-hand and right-hand C–H bonds are in parallel, so now this gives rise to a phenomenon known as **torsional strain**, whereby the electron pairs in the two filled σ orbitals repel each other. This combination raises the energy of the system and makes the adoption of the eclipsed conformation unfavourable. Of course, this effect is tripled because there are three eclipsing sets of parallel C–H bonds (shown in Figure 4.14).

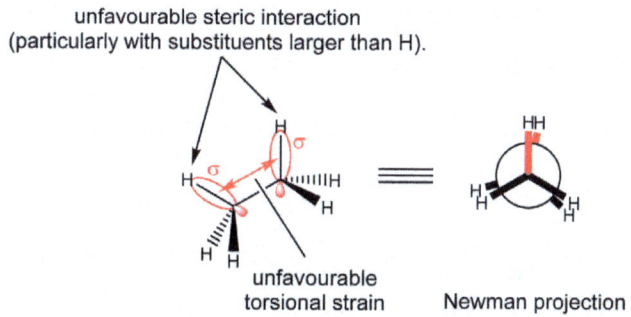

Figure 4.14   Torsional strain in the eclipsed conformation of ethane.

When we rotated the model of ethane about the central C–C bond, we only considered the staggered (minimum-energy) and eclipsed (maximum-energy) conformations, but there are, of course, many conformations in between. Added to this, we have only considered rotation through 60° and not 360°. But because all the substituents in ethane are the same (hydrogen), if we carry on rotating, the symmetry means we will find conformations we have already seen. In fact, there are three staggered and three eclipsed conformations. This is most evident from a plot of the relative conformational energy against the dihedral angle of ethane over 360°, with respect to the two hydrogens shown in red (Figure 4.15).

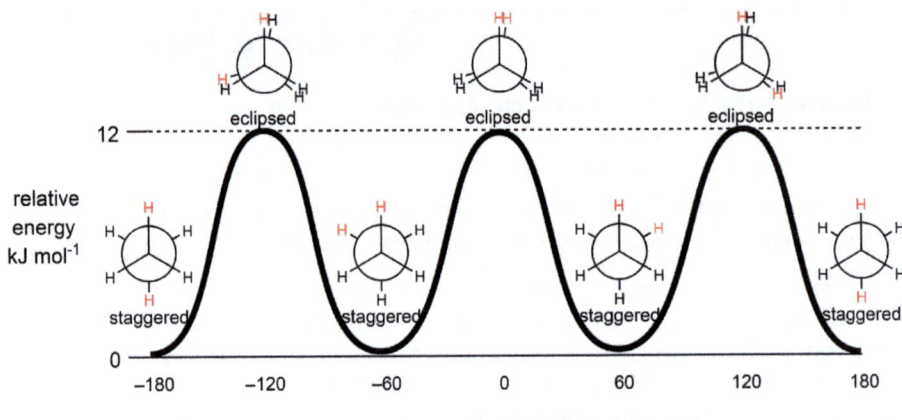

Figure 4.15   Conformational energies of ethane as a function of dihedral angle.

Free rotation[†] can occur when the conformational barrier to rotation is typically below ≈100 kJ mol⁻¹ at room temperature. Above this, the conformers are regarded as different isomers (see Chapter 6). Therefore, for ethane, free rotation is easily achieved and in fact occurs at a frequency of ≈10^{11} s⁻¹. This relatively low barrier to rotation means that, at room temperature, an equilibrium of all conformations can be established for ethane. However, even a relatively small difference in energy still means that the vast majority of molecules (≈99% for ethane) adopt a staggered conformation.

Ethane is a very simple example, since all substituents are hydrogen atoms. The picture gets more complicated when considering more complex molecules, with the following factors often playing a significant role:

• Steric strain (see ethane *vs.* butane in Figure 4.16)
• Torsional strain (see Figure 4.14)
• Hyperconjugation (see Figures 4.11 and 4.17)
• Dipole–dipole interactions (see Figures 4.11 and 4.17)
• Hydrogen bonding (see the exercises at the end of this chapter).

If we consider butane (where two Hs of ethane are replaced by methyl groups), plotting the energy profile (dashed) against that of ethane (solid), we can see that the three staggered and three eclipsed conformations no longer have equivalent energies (Figure 4.16). For the high-energy eclipsed conformations known as the *syn*-periplanar conformations, the two methyl groups eclipse one another. We get a higher level of steric strain because the methyl group is much bigger than H and so the *syn*-periplanar conformation is ≈3 kJ mol⁻¹ higher in energy than the other eclipsed conformation, where Me eclipses H (known as **anticlinal**). For the lower-energy staggered conformations, we find that the conformation with the methyl groups at 60° to one another (known as **gauche** or **synclinal**) is ≈3.8 kJ mol⁻¹ higher

Typically:

• A difference in the energy barrier to rotation of ≈6 kJ mol⁻¹ changes the rate of rotation (at room temperature) by about a factor of 10.
• A rotation rate of 1 s⁻¹ equates to an energy barrier of 73 kJ mol⁻¹.
• Rotation in ethane at room temperature is too fast to measure, but is estimated to be on the picosecond timescale.

[†]Free other than a small barrier to rotation.

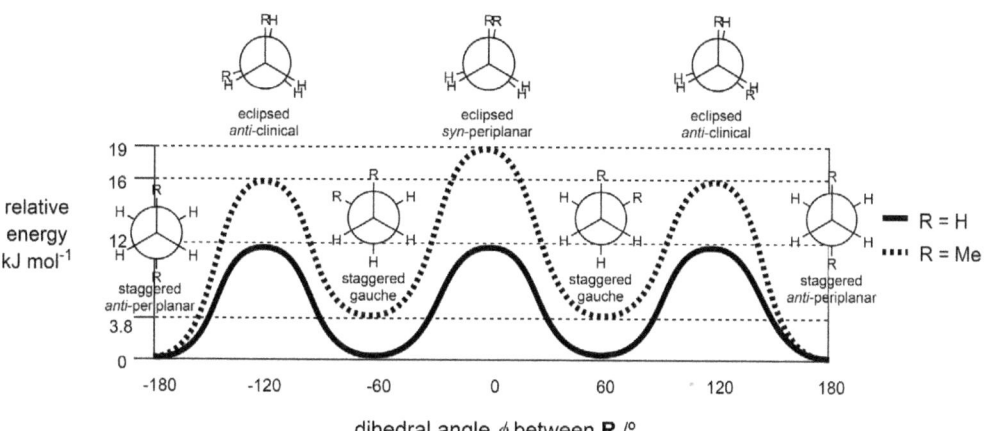

Figure 4.16 Conformational energies of ethane and butane as a function of dihedral angle.

in energy than when the methyl groups are 180° to one another. This, again is due to the greater steric repulsion experienced when the methyl groups are in closer proximity. Although, compared with ethane, butane has a higher barrier to rotation, it is still relatively low and so butane, as with ethane, exhibits relatively free rotation with a rate of $\approx 2 \times 10^9$ s^{-1}, fifty times slower than ethane.

There are examples where the gauche conformation can be lower in energy than the *anti*-periplanar conformation. One such example is 1,2-difluoroethane, shown in Figure 4.17. The lower-energy σ* orbital of the C–F bonds (relative to the C–H bonds) and the higher energy σ orbital of the C–H bonds (relative to the C–F bonds) makes the C–H–σ, C–F–σ* hyperconjugation interaction more favourable and means that the fluorine atoms prefer to sit gauche to one another rather than *anti*-periplanar.

> Head back to Section 2.1.2 if you want a reminder on why the presence of π orbitals (e.g. in alkenes) mean rotation around C=C essentially doesn't occur.

❗ Key Learning Point

When two substituents on adjacent sp³ carbons are in the same plane as each other and as the single bond between the carbons, we say they are **periplanar**. *Syn* means on the same side and *anti* means on the opposite side. So *syn*-periplanar just means the substituents are on the same side in the same plane and *anti*-periplanar means they are on the opposite side in the same plane.

If the substituents we are looking at are not periplanar, they will either be anticlinal, if they are eclipsed, or synclinal, if they are staggered. The staggered synclinal conformation is an important one and, confusingly, has another name, gauche.

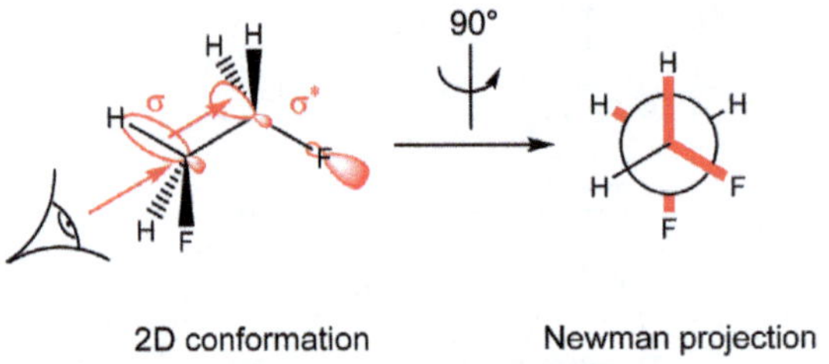

2D conformation   Newman projection

Figure 4.17   The gauche effect of 1,2-difluoroethane.

Let's check our understanding. Looking down the bond as shown by the eye, predict the highest- and lowest-energy conformer for the molecule shown in Figure 4.18. Use a sawhorse projection if you need to.

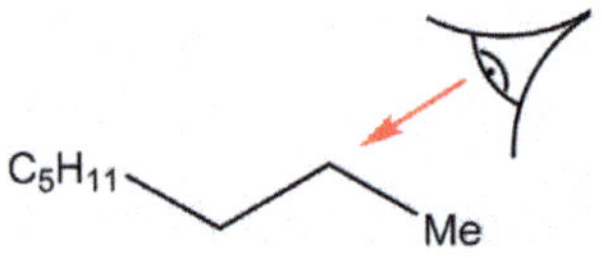

Figure 4.18   Checkpoint example for conformational energies.

Hopefully you've got some neatly drawn projections rather like the ones shown in Figure 4.19 and Figure 4.20!

Now take your Newman projection and rotate the back carbon through 180°. Well done. You've made the highest energy conformation (Figure 4.20).

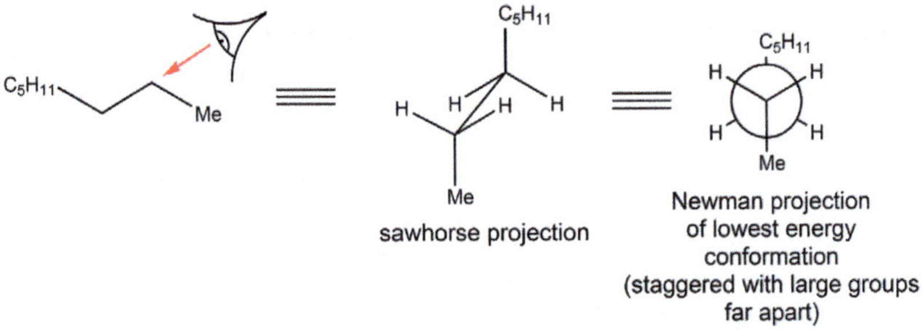

sawhorse projection

Newman projection of lowest energy conformation (staggered with large groups far apart)

Figure 4.19   Checkpoint example for conformational energies.

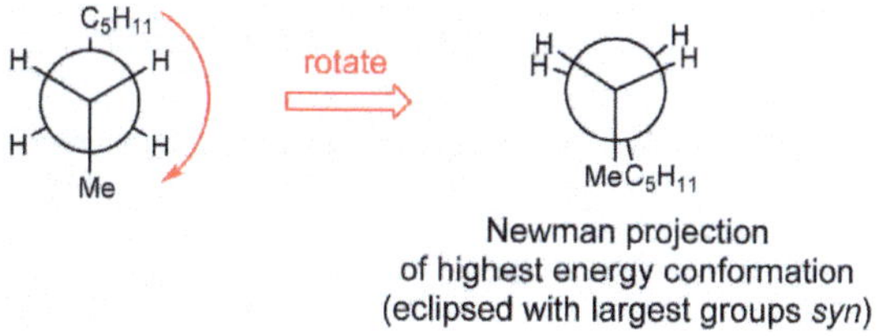

Newman projection
of highest energy conformation
(eclipsed with largest groups *syn*)

Figure 4.20   Checkpoint answers for conformational energies.

4.5   Relative Stereochemistry

In Chapter 3, we met the term absolute stereochemistry, *i.e.* the precise spatial arrangement of the substituents around a stereogenic centre. It can also be very useful to describe the spatial arrangement of one group *relative to* another group in a diastereoisomer; this is known as **relative stereochemistry**. In fact, we have already met this idea in Section 2.3.1 with alkenes, where the *cis* and *trans* prefixes can be used to describe the position of two identical alkene substituents (*e.g.* H or methyl groups) *relative to one another e.g. cis-* and *trans-*but-2-ene (Figure 4.21).

cis-but-2-ene   *trans*-but-2-ene

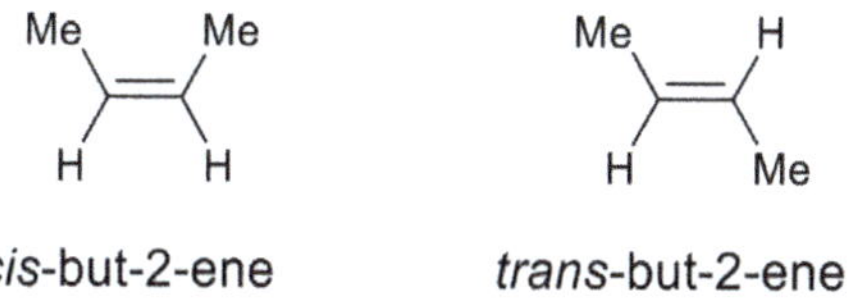

Figure 4.21   *Cis-* and *trans-*but-2-ene (see also Figure 2.10).

---

❶ Key Learning Point

Relative stereochemistry refers to the spatial arrangement of a substituent (*e.g.* at one stereogenic centre) relative to a substituent at another atom.

Since (*E*)- and (*Z*)-alkenes cannot be interconverted without breaking the C=C bond, they can be considered diastereoisomers, specifically achiral diastereoisomers. It follows then that in the same way that we use *cis* and *trans* to describe the relative position of two groups in an alkene (Figure 4.21), we can use the same convention in other molecules where the two substituents' relative positions cannot be changed through bond rotation. This is particularly common in substituents on cyclic compounds, such as those depicted in Figure 4.22. Here the relative stereochemistry of two specified (circled) substituents is described, but if no particular substituents are specified, we can only use the *cis* and *trans* prefixes if both of the following apply:

Be careful not to confuse *conformation*, the relationship between isomers interconverted *via* the rotation of single bonds, with *configuration*, the relationship between isomers that have different spatial arrangement of atoms in a molecule and can only be altered by breaking and making bonds.

- We are describing two atoms that are stereogenic (*i.e.* they do not have two or more identical substituents).
- We are describing two stereogenic centres that each have one common substituent (*e.g.* H).

(a)

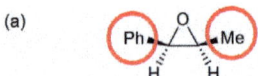 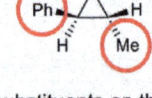

circled substituents on the same side and cannot interconvert without breaking bonds ∴ *relative stereochemistry = cis*

circled substituents on the opposite sides and cannot interconvert without breaking bonds ∴ *relative stereochemistry = trans*

(b)

circled substituents on the same side and cannot interconvert without breaking bonds ∴ *relative stereochemistry = cis*

circled substituents on the opposite sides and cannot interconvert without breaking bonds ∴ *relative stereochemistry = trans*

Figure 4.22   Using *cis* and *trans* with restricted rotation.

Have a go at making models of the two molecules in Figure 4.22a and convince yourself that the two isomers cannot be interconverted through bond rotation. Instead bonds must be broken. Another way of understanding this is to assign the absolute configuration of the two stereogenic centres. The *cis*-epoxide is (*R*,*S*), whereas the *trans*-epoxide is (*R*,*R*). They are **diastereoisomers**.

Then check the same thing with Figure 4.22b. These examples can all be classified as configurational isomers.

If we now make models of the two molecules shown in Figure 4.23, we find that in this case they *can* be interconverted through rotation

 As in Chapter 3, the ∴ symbol is a useful shorthand for 'therefore' which you will see in Figures 4.22, 4.23, 4.26 and 4.27.

circled substituents on the opposite side ∴ *relative stereochemistry = anti*

circled substituents on the same side ∴ *relative stereochemistry = syn*

Figure 4.23   Using *syn* and *anti* with free rotation. The left-hand representation has the Me and OH on opposite sides with the heptane chain drawn in zig-zag, so this isomer is the *anti* diastereoisomer.

about the C–C single bond shown and so are not configurational isomers. Indeed, applying the CIP rules shows they have the same absolute configuration at both stereogenic centres. Instead, these molecules are conformers of one another.

It's clearly useful to be able to describe the stereochemical, 3D, relationship between the methyl and hydroxy substituents in the conformers shown in Figure 4.23. It's not standard practice to use *cis/trans* nomenclature here, because the substituents are not permanently on the same or opposite sides of the single bond. Instead we need to denote the relative stereochemistry. So, is this molecule the *anti* or the *syn* diastereoisomer? In such cases, we draw the longest chain of the molecule in the normal zig-zag representation, as in the left example. We then refer to diastereoisomers with substituents (non-hydrogen ones) on the same face as being *syn* and diastereoisomers with substituents on opposite faces as being *anti*.

An older naming system for the same purpose with adjacent stereogenic centres uses the prefixes *threo* and *erythro* (derived from the carbohydrates threose and erythrose, respectively). This approach has become largely redundant other than in carbohydrate chemistry and so for the purposes of this textbook, we will stick to using **syn** and **anti**.

Consider again the four stereoisomers of ephedrine, discussed in Chapter 3 (Figure 4.24). Now that you understand the concepts of absolute and relative stereochemistry, you have new methods to use to decide whether something is an enantiomer, a diastereoisomer, a conformer or the same molecule:

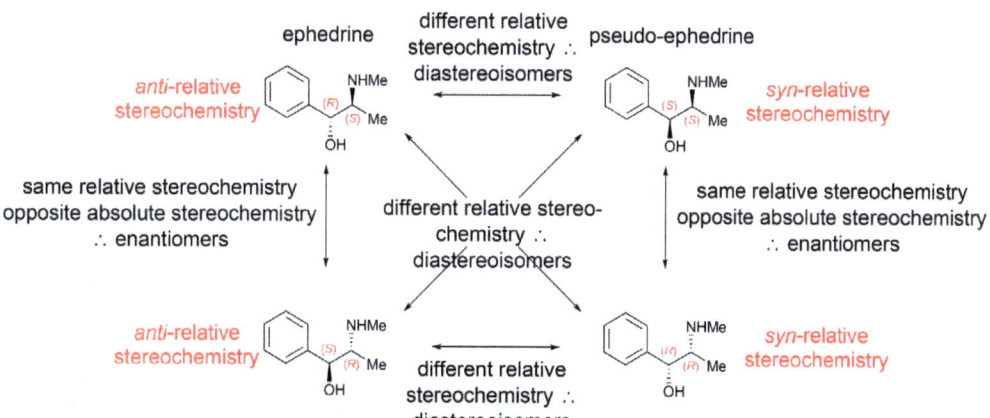

Figure 4.24   Relative and absolute stereochemistry in ephedrine isomers.

- An enantiomer will have the same relative stereochemistry but the opposite absolute stereochemistry (configurations of all stereogenic centres are inverted).
- A diastereoisomer will have different relative stereochemistry and different absolute stereochemistry (configuration if at least one but not all stereogenic centres is inverted).
- A conformer will have the same absolute and relative stereochemistry (when drawn in a zig zag representation) and will have had one or more single bonds rotated.

 Checkpoint

You should now understand the meaning of *anti* and *syn* relative stereochemistry.

Let's check our understanding. What is the relative stereochemistry of the two alcohol groups in the conformation shown in Figure 4.25? Check your answer against Figure 4.26.

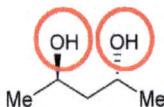

Figure 4.25   Checkpoint example for relative stereochemistry.

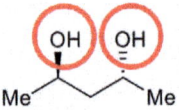

Circled substituents are on the opposite side with the longest chain drawn in zig-zag form ∴ *anti*.

Figure 4.26   Checkpoint answer for relative stereochemistry.

In NMR, the interaction of nuclei through bonds (coupling) gives rise to a coupling constant (*J*), measured in hertz. The scalar magnitude of the coupling increases as a function of the efficiency of the orbital interactions on the pathway between the nuclei. Generally, shorter pathways with better orbital overlap (parallel interactions) give larger coupling constants.

We are assuming a basic level of knowledge about NMR. If you are unsure of any theory discussed here, please go to your spectroscopy textbooks.

4.6 Conformation and NMR

It was mentioned earlier that energy barriers to rotation above ≈100 kJ mol⁻¹ at room temperature lead to such a slow rate of interconversion that conformers of a molecule are regarded as different isomers. A typical example of this is with alkenes, discussed in more detail in Chapter 2, which typically have a room temperature energy barrier to rotation in the region of ≈260 kJ mol⁻¹. This gives rise to *E/Z* isomers that can be separated and, importantly, exhibit different physical and spectroscopic properties. One such property is the ¹H NMR spectra, in particular, the ³*J* vicinal coupling constants of alkene protons, which can be used to determine the geometry of an alkene. Consider the two stereoisomers of pent-2-ene, *E* and *Z*, shown in Figure 4.27. The ¹H NMR ³*J* values between the protons depicted in red are 11.0 Hz and 15.0 Hz for the *cis* and *trans* protons, respectively. This is typical of alkenes, with almost no exceptions to the general rule that for a pair of geometric alkene isomers, ³*J(trans)* > ³*J(cis)*. Furthermore, although there can be some overlap to *J* value ranges (especially with particularly electronegative or electropositive adjacent atoms), we find that, in general, *cis*-alkene protons exhibit vicinal *J* values in the range of 7–11 Hz (but these can be as low as 4 Hz for alkenyl fluorides and as large as 19 Hz for alkenyl lithiums) and *trans*-alkene protons exhibit *J* values in the range of 14–18 Hz (but these can be as low as 12 Hz for alkenyl fluorides and as large as 24 Hz for alkenyl lithiums) (see Chapter 2).

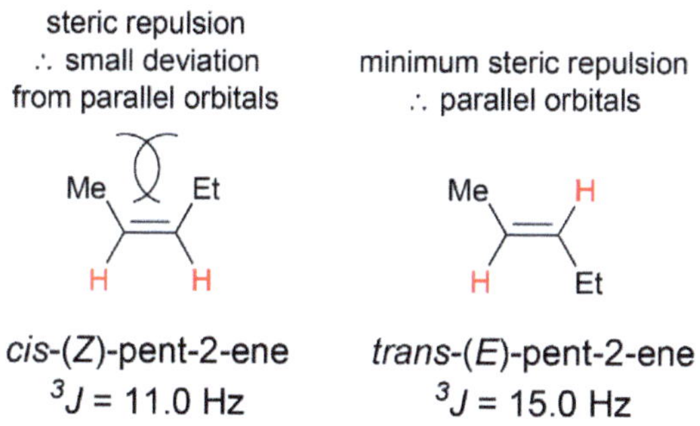

Figure 4.27   *Cis*- and *trans*-pent-2-ene.

> ⓘ **Key Learning Point**
>
> *Vicinal* comes from the Latin *vicinus*, meaning neighbour. The terms **vicinal angle** and **dihedral angle** ϕ mean the same thing, but NMR spectroscopists tend to say vicinal.

The reason behind this difference in the ^{1}H NMR coupling constant stems from the orbitals in the system. Coupling is a through-bond phenomenon and so orbitals play a fundamental role. As discussed previously with hyperconjugation, parallel orbitals have the greatest interaction and this gives rise to the variation in coupling constant magnitude: those bonds whose orbitals have a greater parallel alignment will exhibit a larger scalar coupling. The Karplus equation models this (ϕ = dihedral angle and *A*, *B* and *C* are parameters that depend on the nature of the molecule). Plotting the equation as the Karplus curve (Figure 4.28, values vary depending on the system in question) gives a quick way to predict coupling constants between protons at a particular dihedral angle to each other.

$$J(\phi) = C\cos^2\phi + B\cos\phi + A$$

As can be seen, where the dihedral angle is 0° or 180° (so the protons are periplanar), the Karplus curve predicts that the vicinal coupling constant *J* is at its greatest. The value is larger at 180° because more of a parallel orbital alignment can be achieved when the larger substituents (R ≠ H) are further apart than when

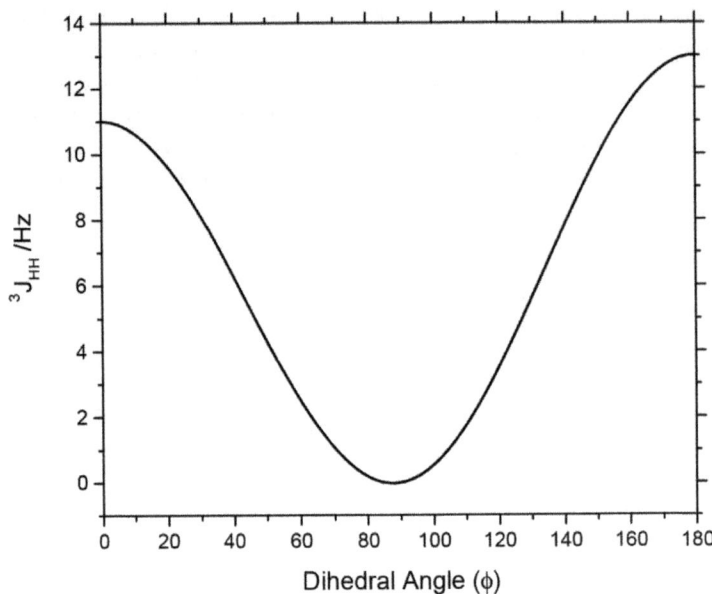

Figure 4.28 The Karplus curve.

the larger substituents are closer together and hence experience more steric repulsion, *i.e.* they deviate from true parallel by a small amount (similar to Figure 4.27). When the bonds (and hence the orbitals) are perpendicular (90°), *J* is predicted to be at its smallest. For saturated alkyl chains, such as ethane and butane, with relatively free rotation about C–C bonds, as discussed previously, the vicinal couplings ^3J tend to average out at around 7 Hz. This relationship between the vicinal (dihedral) angle and coupling constant allows us to estimate the conformation that molecules adopt, particularly those with restricted rotation, such as alkenes and cyclic alkanes (see Chapter 5).

Conformation not only affects the magnitude of coupling constants in NMR spectroscopy, but also the chemical shift and the equivalence of different nuclei. Consider the common solvent *N*,*N*-dimethylformamide, shown in Figure 4.29. On the face of it, the two methyl groups would appear to be equivalent if we assume free rotation about the C–N single bond. However, the resonance effect of N lone-pair donation into the C=O π* orbital means that the C–N bond has some double bond character and, as with alkenes, experiences restricted rotation at room temperature, with an energy barrier of ≈88 kJ mol⁻¹. This means that the methyl group ¹H nuclei are in different environments and hence exhibit different chemical shifts, giving rise to two separate peaks, at ≈3.0 and ≈2.9 ppm. This is because NMR is a fast enough technique to be able to 'see' both methyl groups in this case, rather than an average. The 'NMR timescale' is around 1000 s^{-1}, or a rotational energy barrier around 55 kJ mol^{-1}, depending on the nucleus being observed and the spectrometer being used. So, what would happen if we ran the ^{1}H NMR spectrum of *N,N*-dimethylformamide at higher temperatures?

N,N-dimethylformamide

Figure 4.29   Restricted rotation in partial double bonds *e.g. N,N*-dimethylformamide (DMF).

That's right, if a variable temperature NMR experiment is run, as we heat the sample the rotational energy barrier becomes less significant relative to the NMR timescale. The two ^{1}H NMR peaks for the methyl groups coalesce into a single peak when we reach around 90–100 °C.

4.7   Stereochemistry and NMR

As discussed in Chapter 3, enantiomers exhibit identical chemical, physical and spectroscopic properties other than their optical rotation, while diastereoisomers have distinct physical and spectroscopic properties and can also undergo different chemical reactions. Consider the molecules 3-hydroxypropionic acid and L-serine, shown in Figure 4.30. You might think that the nuclei in red would be equivalent by ^{1}H NMR for each molecule. For 3-hydroxypropionic acid we find that the protons in red are indeed equivalent and hence only a single 2H peak, around 3.8 ppm, is observed. For L-serine, however, two different ^{1}H signals, around 4 ppm, are observed and the two protons in fact couple to each other with a large **geminal** ^2J value of about 13 Hz. This is because they are in fact inequivalent, **diastereotopic** protons.

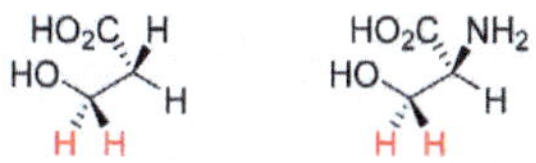

3-hydroxy-
propionic
acid

L-serine

Figure 4.30   Equivalent and diastereotopic protons.

> **◼Checkpoint**
>
> You should now understand how conformation influences coupling constants in ^{1}H NMR spectroscopy and be able to predict the coupling constant by considering the vicinal/dihedral angle between two proton nuclei.

To explain what this means, for both molecules, we will label one of the red hydrogens H_a and the other H_b. Firstly considering 3-hydroxypropionic acid (Figure 4.31), if we were to swap H_a for a different group R (R ≠ H) and then separately swap H_b for an R group and then compare the two molecules, we find that we have a pair of enantiomers because we have created a new stereocentre. The two red protons are therefore considered to be **enantiotopic**. We know already that enantiomers cannot be distinguished by ^{1}H NMR and, similarly, we find that enantiotopic protons are not distinguishable by ^{1}H NMR. They are therefore considered to be **magnetically equivalent.**

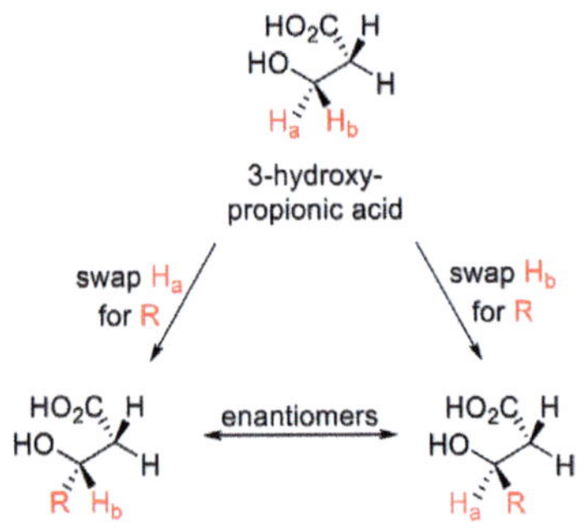

Figure 4.31   Enantiotopic protons.

Now have a go at carrying out the same exercise with L-serine. Swap one red H for R and then the other (Figure 4.32). What is the stereochemical relationship between the two new compounds you have drawn? Would the two red protons have the same chemical shift then?

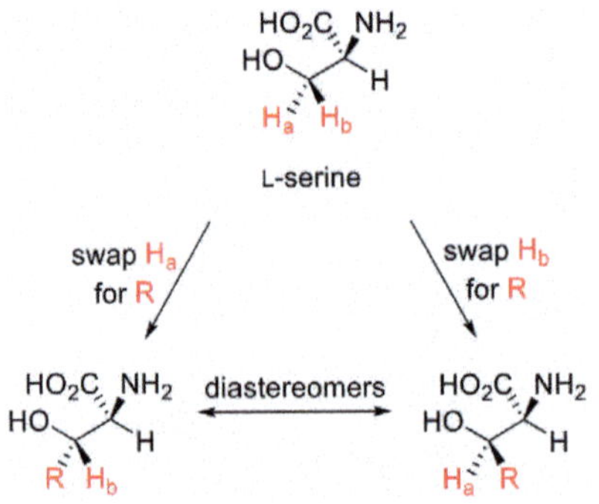

Figure 4.32   Diastereotopic protons.

> **❶ Key Learning Point**
>
> Diastereotopicity refers to a relationship between two groups in a molecule where replacement of one of the two groups *vs.* the other gives rise to a pair of diastereoisomers. Similarly, enantiotopic is where replacement of one group *vs.* the other group gives enantiomers (see 3-hydroxypropionic acid or the amino acid glycine) and **homotopic** is where replacement of one group *vs.* the other gives the same compound (see chloromethane).

The two new compounds with an R group have the same configuration at one stereogenic centre and the opposite at the other. Check by assigning (*R/S*) if you can't see it straight away. So, we now obtain a pair of diastereoisomers. The two red protons are therefore considered to be diastereotopic. We know already that diastereoisomers are distinguishable by ^{1}H NMR spectroscopy and, similarly, we find that diastereotopic protons are distinguishable by ^{1}H NMR.

So, H_a and H_b in L-serine have different chemical shifts in the ^{1}H NMR spectrum.

> **◼ Checkpoint**
>
> You should now understand what is meant by the terms enantiotopic and diastereotopic and how this property affects ^{1}H NMR spectra.

4.8 Conclusion

We've covered a number of challenging concepts in this chapter, such as configuration *vs.* conformation, absolute *vs.* relative stereochemistry and how NMR spectroscopy can be used to study stereochemistry. We've also learned how to draw Newman projections and use them to predict relative energies of different conformations. Well done! You'll find that all these concepts are developed further in Chapter 5, when we start looking at cyclic compounds, such as cyclohexanes and sugars. So there will be plenty more chance to practise, but don't forget to look back to this chapter for reference when you need to.

Exercises

Questions 1, 2, 4 and 5 are quick checks of your understanding, though they might form simple openers to exam questions. Questions 3, 6 and 7 are more typical of parts of exam questions.

1. For each of molecules (i) to (iii), draw a Newman projection of the conformation shown, from the perspective of the eye.

2. For the Newman projections shown, convert the molecules to the corresponding 2D depiction (as in Question 1), putting the front carbon atom on the *right*-hand side.

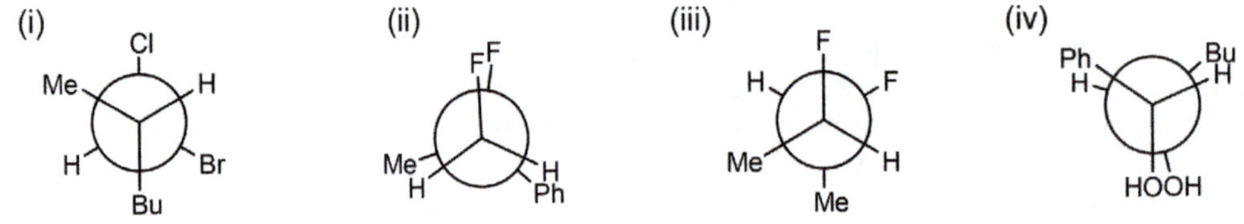

(i)

(ii)

(iii)

(iv)

3. Predict the highest- and lowest-energy conformer for the molecules shown, bearing in mind that formation of intramolecular hydrogen bonds can be very important energetically.

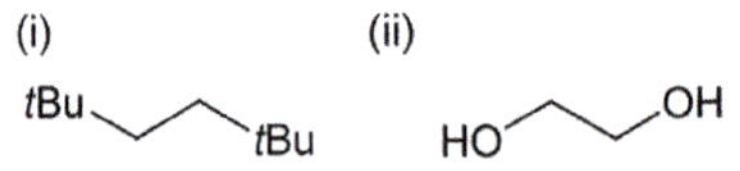

(i)

(ii)

4. Assign the relative stereochemistry as *cis* or *trans* with respect to the circled substituents in the molecules shown.

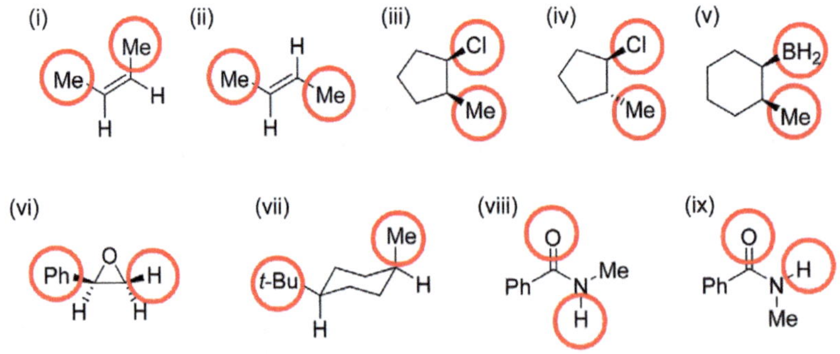

(i)

(ii)

(iii)

(iv)

(v)

(vi)

(vii)

(viii)

(ix)

5. Assign the relative stereochemistry as *syn* or *anti* with respect to the circled substituents in the molecules shown.

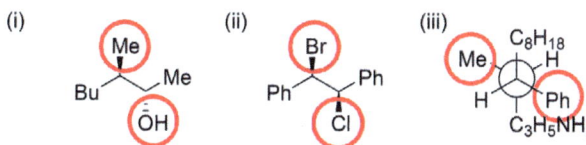

6. Indicate how many ^{1}H NMR signals are expected for the protons at the positions indicated in the molecules shown.

7. Consider the antidepressant drug fluoxetine (Prozac) shown in Figure 4.33 and answer the following questions:
    i. Draw the *R* enantiomer of fluoxetine.
    ii. Draw a Newman projection for the highest- and lowest-energy conformers for the *R* enantiomer from the viewpoint of the eye shown in Figure 4.33 (assume that Ph is more sterically demanding than the aryloxy group).
    iii. Draw the *S* enantiomer of fluoxetine.
    iv. Draw a Newman projection for the highest- and lowest-energy conformers for the *S* enantiomer from the viewpoint of the eye shown in Figure 4.33 (assume that Ph is more sterically demanding than the aryloxy group).
    v. Would you expect the CH$_2$ group adjacent to the stereogenic carbon to exhibit one or two ^{1}H NMR signals? Why?

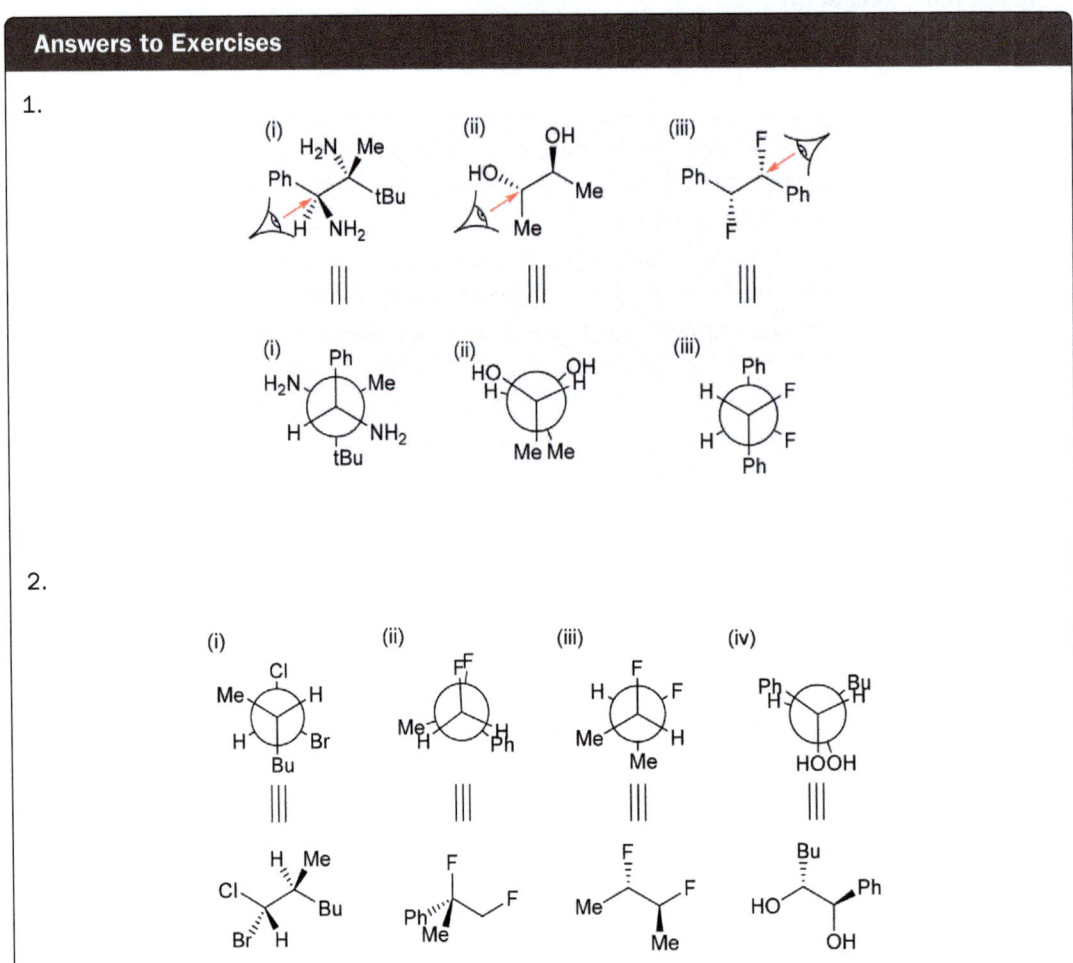

fluoxetine (Prozac)

Figure 4.33   Fluoxetine.

Answers to Exercises

3.

(i)

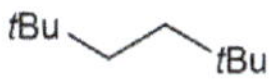

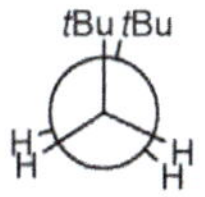

tBu ~~~ tBu ≡

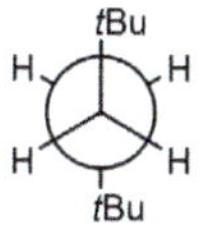

Newman projection
of lowest-energy
conformation
(staggered with large groups
far apart)

Newman projection
of highest-energy conformation
(eclipsed with largest groups *syn*)

(ii)

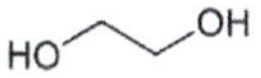

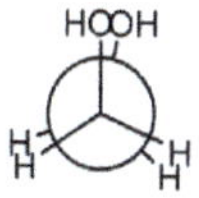

HO ~~~ OH ≡

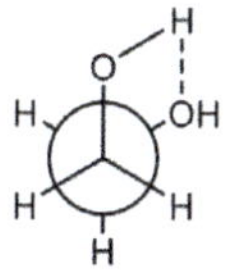

Newman projection
of lowest-energy
conformation
(OH groups *gauche* -
intramolecular H bond)

Newman projection
of highest-energy
conformation (eclipsed OH
groups)

4.

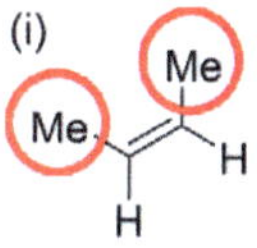

circled substituents
together, so
cis

(i)

(ii)

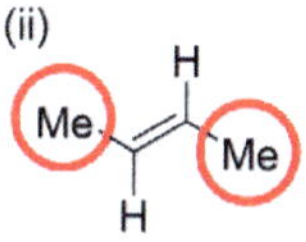

circled substituents
opposite, so
trans

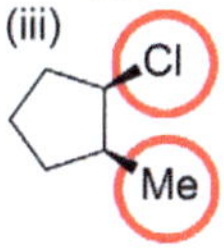

circled substituents
together, so
cis

(iii)

(iv)

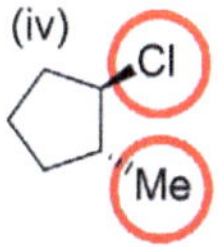

circled substituents
opposite, so
trans

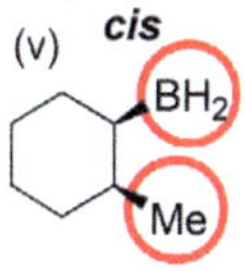

circled
substituents
together, so
cis

(v)

(vi)

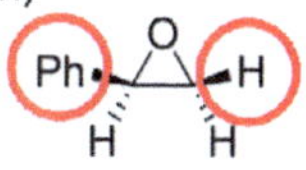

circled substituents
together, so
cis

(vii)

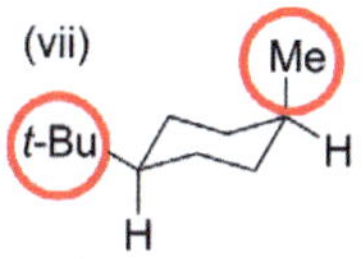

circled substituents
together, so
cis

see Chapter 5 for help

(viii)

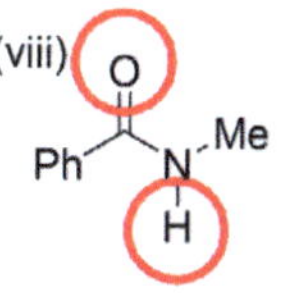

circled substituents
opposite, so
trans

(ix)

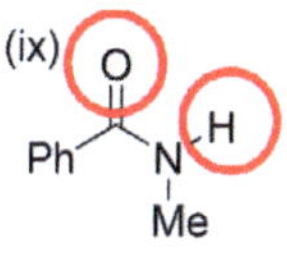

circled substituents
together, so
cis

Remember that amide C–N bonds have
partial double bond character, so
restricted rotation.

5.

(i)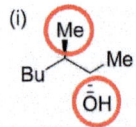

circled substituents
opposite, so
anti

(ii)

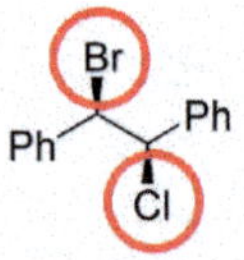

circled substituents
together, so
syn

(iii) C_8H_{18}
Me, H
H, Ph
$C_3H_5NH_2$

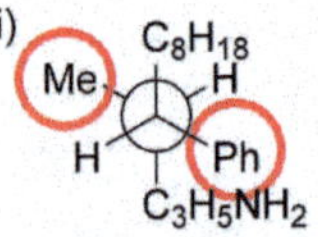

circled substituents
opposite, so
anti
(*importantly, anti-
periplanar*)

6.

(i) O
Me, N, Me
Me

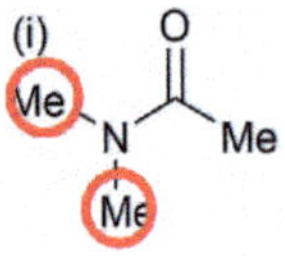

2 separate signals (at room
temperature). Partial C–N
double bond character leads
to restricted rotation.

(ii) O
Me, O

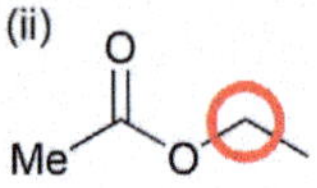

1 signal – no significant
restricted rotation and no
stereogenic centre ∴ Hs are
equivalent.

(iii)
Ph, OMe

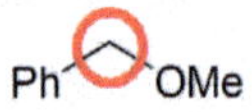

1 signal – no
stereogenic centre ∴
Hs are equivalent.

(iv) OH
*

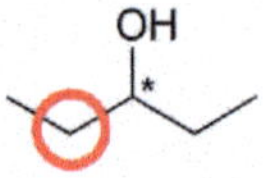

2 signals – adjacent
stereogenic centre * ∴
Hs are diastereotopic
and inequivalent.

(v)
H_2N, OMe
O

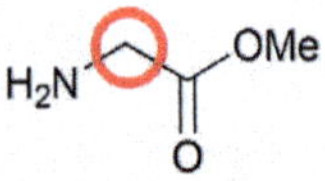

1 signal – no
stereogenic centre
∴ Hs are
equivalent.

(vi)
H_2N, OH
*
O

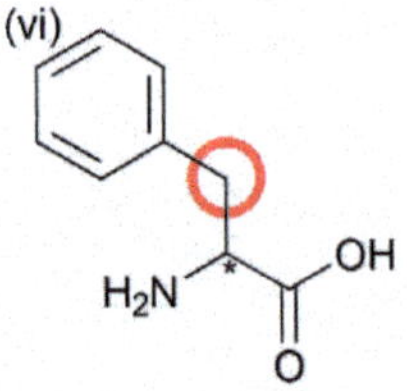

2 signals – adjacent
stereogenic centre * ∴
Hs are diastereotopic
and inequivalent.

(vii)
*, NH_2

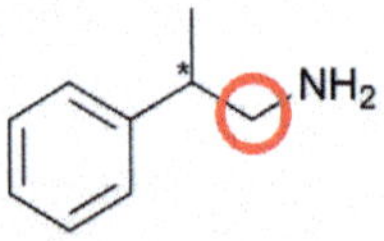

2 signals – adjacent
stereogenic centre * ∴
Hs are diastereotopic
and inequivalent.

(viii) H
H_2C, Bu

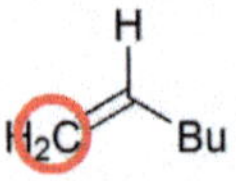

2 separate signals (at
room temperature).
Double bond leads to
restricted rotation.

7.

i.

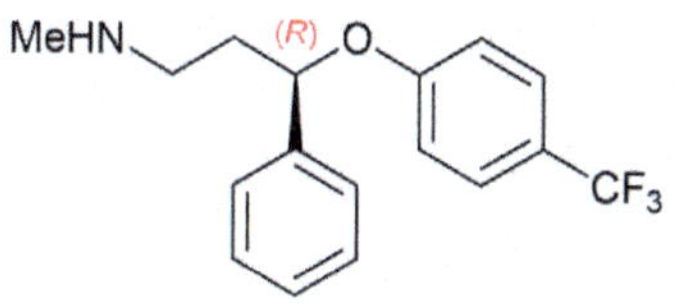

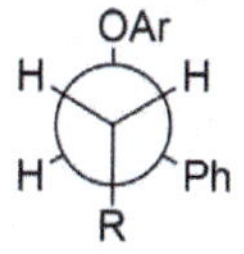

ii. See (i)

iii.

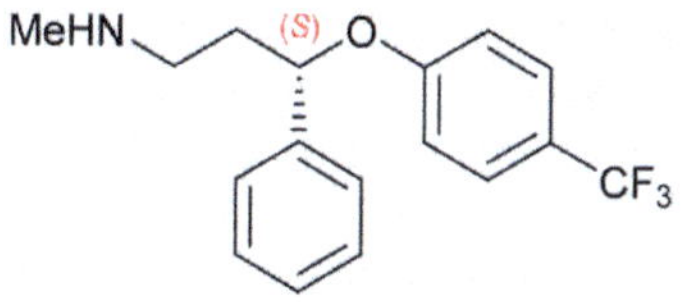

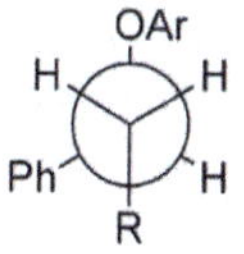

iv. See (iii)

v. We would expect two separate ^{1}H NMR signals because the protons are diastereotopic, owing to the adjacent stereogenic carbon.

By the End of This Chapter You Will:

- ☐ Understand the conformational analysis of six-membered rings and of other ring sizes.
- ☐ Be able to draw conformations for cyclohexanes that are part of more complex structures.
- ☐ Be able to accurately represent six-membered rings and their substituents in 3D.
- ☐ Be able to apply conformational analysis in ^{1}H NMR interpretation.

What You Will Get from This Chapter

In most of the compounds we've studied so far, rotating around a bond has a relatively minor effect on the rest of the molecule. When we look at ring structures it's different. The geometrical constraints mean that a change in one place will have a big impact on other carbon atoms. Key to understanding this stereochemistry is the concept of 'ring flip'. It may look at first sight as though we've just turned a molecule round or over, but in fact ring flip is an important change in conformation. If you use your molecular models and take time to understand, you'll really enjoy this fascinating chemistry.

Conformation of Cyclic Compounds

Natural molecules often contain six-membered rings and, of these, sugars are arguably the most important. They have many different roles in biology; from their use as an energy source, to forming shells of crustaceans and even helping your body fight diseases. You've probably heard of glucose, but there are many other types of sugars, such as galactose and mannose. Can you spot the differences in their structures (Figure 5.1)?

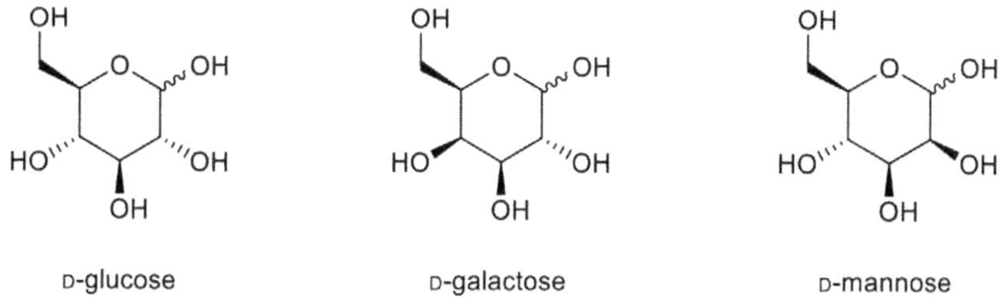

D-glucose   D-galactose   D-mannose

Figure 5.1 Sugars are cyclic molecules.

Only the stereochemistry of one or two stereogenic centres is different in each structure, meaning that these sugars are diastereoisomers (diastereoisomers were introduced in Chapter 3). However, such apparently small changes result in these sugars behaving very differently in your body. To understand how they function, we need to think about these molecules in 3D.

In this chapter, we will explore the conformations of ring structures. We will focus on six-membered rings, but will briefly discuss other ring sizes. We will also consider what happens to groups attached to the ring and when sp^2 hybridisation (Chapter 2) is present inside and outside of the ring. Finally, this chapter focuses on how to apply conformational analysis of cyclic systems in practice, namely how to draw these molecules in 3D and correctly assign their ^{1}H NMR spectra.

5.1 Conformation of Six-membered Rings

5.1.1 Saturated Six-membered Rings Have Chair Conformations

We are going to start by comparing two of the simplest six-membered rings in organic chemistry, benzene and cyclohexane. If we view both molecules from above, they are hexagonal. However, when we rotate the structure to view the molecules from the side, we can see the effect the different carbon hybridisations have on their shapes.

Every carbon of benzene (Figure 5.2) is sp^2 hybridised, meaning that each carbon is trigonal planar and from the side it appears flat, which it is.

benzene

top view: side view:

all carbons are sp^2 hybridised   planar

Figure 5.2 Representations of benzene.

For cyclohexane (Figure 5.3), all carbons are sp^3 hybridised. The arrangement of bonds around each carbon is tetrahedral, which means that the shape of the ring cannot be planar. The shape of a cyclohexane is called a chair conformation (Figure 5.4), because it resembles a deck chair (at least to the chemist who gave this conformation its name).

cyclohexane

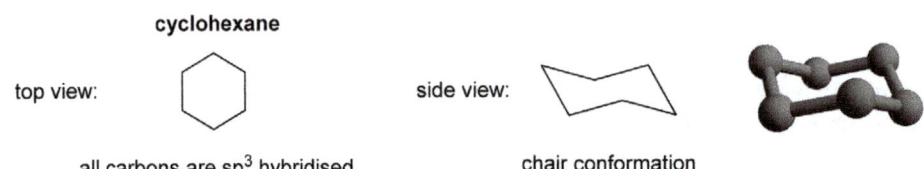

top view: side view:

all carbons are sp^3 hybridised   chair conformation

Figure 5.3 Representations of cyclohexane.

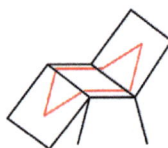

Figure 5.4 The conformation of cyclohexane resembles a deck chair.

Chairs are the most stable conformation that cyclohexanes can adopt but they can also form another conformation where the two end (as we've drawn them) carbons are both on the same side of the other four carbons, rather than on opposite sides. This is called a **boat conformation** (Figure 5.5).

boat conformation

Figure 5.5   Boat conformation of cyclohexane.

The carbons of both chairs and boats are tetrahedral but to understand why a chair is more stable than a boat it helps to observe their Newman projections (these were introduced in Chapters 1 and 4). All the C–H bonds of a chair conformation are staggered, whereas there are four sets of eclipsing C–H bonds in a boat conformation (Figure 5.6). Remember from Chapter 4 that staggered conformations are lower in energy than eclipsing conformations.

> A simple way to remember the difference between chair and boat conformations is that chairs have the 'pointy ends' facing in opposite directions whilst boat conformations have the 'pointy ends' facing in the same direction.

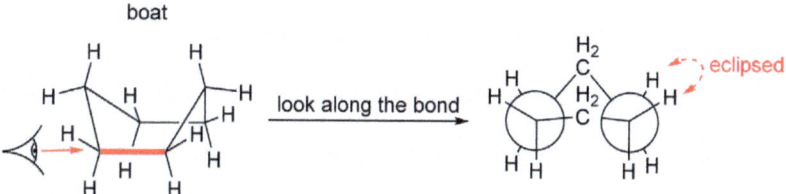

Figure 5.6   Newman projections of chair and boat conformations.

> Although the boat conformation is possible, it is an energy maximum. In contrast to the chair conformation, the boat doesn't contribute much to cyclohexane chemistry.

Let's now check our understanding of the conformations that cyclohexanes can adopt and their stability.

Of the three conformations in Figure 5.7, which can methyl cyclohexane adopt and which is the most stable?

planar conformation   chair conformation   boat conformation

Figure 5.7   Conformations of cyclohexane.

As we said, every carbon in methyl cyclohexane is tetrahedral, meaning that it cannot adopt a planar conformation. However, it can form a chair or a boat conformation. Of the two possible conformations, the chair is more stable as it avoids unfavourable eclipsing interactions.

> **❗ Key Learning Point**
>
> The chair conformation of cyclohexane is the lowest-energy conformation, more stable than the boat or other possible conformations.

5.1.2 Substituents Adopt Either Axial or Equatorial Positions

If we add the hydrogens to the chair conformation (Figure 5.8), we can see that the hydrogens are pointing in different directions relative to the ring.

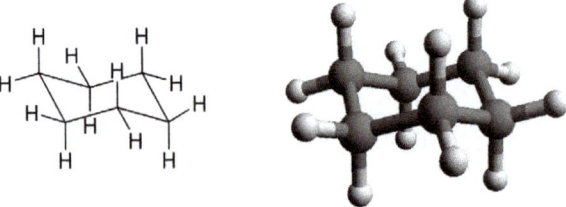

Figure 5.8   Hydrogens on cyclohexane.

Some hydrogens point either straight up or straight down above and below the ring. These are called **axial** hydrogens. If we look at these axial hydrogens more closely, we can see that they alternate between pointing up and pointing down as we go around the ring (Figure 5.9).

axial hydrogens point
above and below the ring

equatorial hydrogens point
to the side of the ring

Figure 5.9   Axial and equatorial hydrogens on cyclohexane.

Other hydrogens point to the side of the ring. These are called **equatorial** hydrogens. In fact, each carbon has one axial hydrogen and one equatorial hydrogen. As you move around the ring, the type of hydrogen that is on the top and bottom alternates between being axial and equatorial.

5.1.3 Chairs Can Ring Flip (Invert)

So far, we've introduced one form of chair conformation, where the carbon on the left is pointing up and the carbon on the right is pointing down. However, another chair conformation can exist where the left and right carbons point in opposite directions. These two chair conformations can actually interconvert with one another in a process called ring flipping (Figure 5.10).

Imagine the chair conformations superimposed on a globe. The axial hydrogens will point towards the axes (*i.e.* north and south pole) and the equatorial hydrogens will point towards where the equator will be.

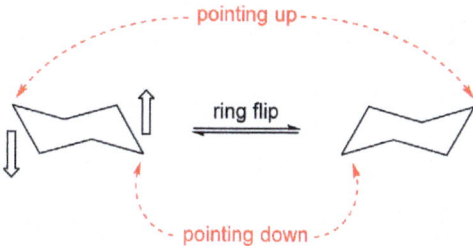

Figure 5.10 Structures of chairs before and after a ring flip.

The cause of ring flipping is rotation around carbon–carbon bonds, but in a ring, this results in the inversion of one chair to another.

A ring flip starts with two of the carbons (labelled 2 and 3) inverting to have a linear configuration with their neighbouring carbons, known as a **half-chair** conformation. Carbons 2 and 3 continue their inversion and end up pointing in the opposite directions to form a **twist-boat** conformation. Next, the two carbons on the opposite side of the ring (5 and 6) start to invert, going through another half-chair conformation before ending up as the flipped chair (Figure 5.11).

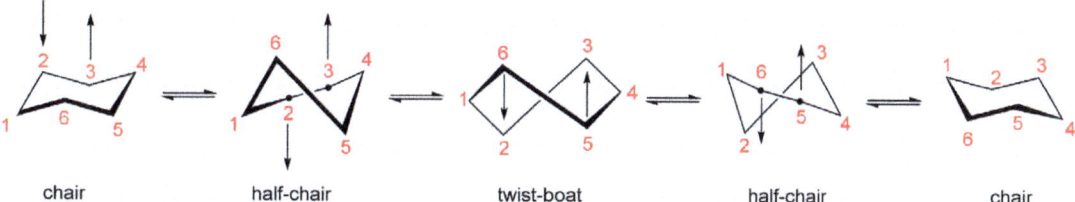

Figure 5.11 How a chair conformation can ring flip to an inverted chair.

Ring flipping also has an effect on the axial and equatorial configuration of the substituents. Groups that are axial in the original chair become equatorial after a ring flip, and equatorial groups become axial (Figure 5.12).

> **❗ Key Learning Point**
>
> Groups that are axial in the original chair become equatorial after a ring flip, and equatorial groups become axial.

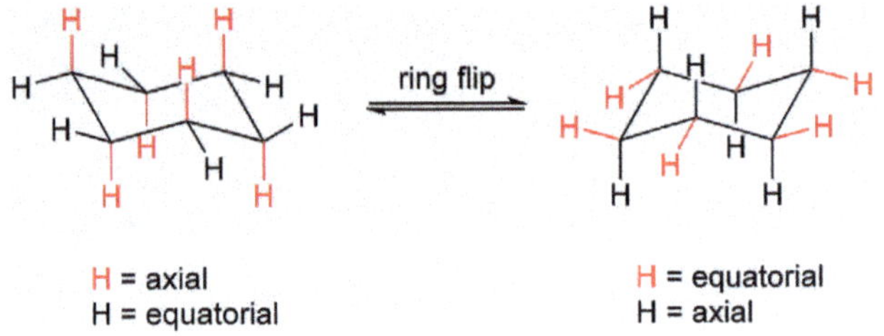

Figure 5.12   How axial and equatorial groups interconvert from a ring flip.

Notice how we use an equilibrium arrow to represent the inter-conversion of the two chairs. There is an energy barrier to forming a half-chair and twist-boat conformation (45 kJ mol⁻¹) but at room temperature there is enough energy for the ring flipping to occur rapidly. In the next section, we see how the equilibrium distribution is altered when we have substitution around the ring.

5.1.4   Substituted Cyclohexanes

Things get a little more complicated when you substitute hydrogens for other groups. Let's have a look at look what happens when we exchange one of the hydrogens for a methyl group. In one chair conformation, the methyl group is equatorial, and when the chair flips it becomes axial (Figure 5.13).

💬

Remember that axial becomes equatorial and *vice versa* after a ring flip.

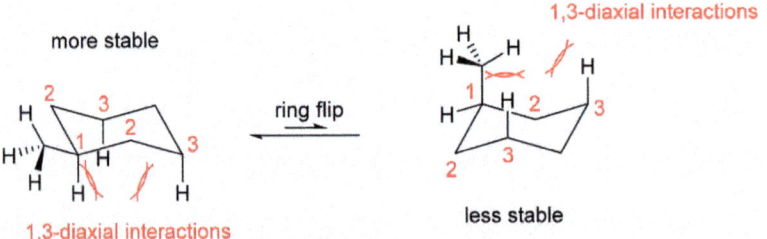

Figure 5.13   1,3-diaxial interactions with axial substituents.

Axial substituents have unfavourable **1,3-diaxial interactions** with other axial groups on the same side of the ring. These are caused by steric and torsional strain. Although all axial groups experience 1,3-diaxial interactions, the larger methyl group experiences greater steric and torsional strain than a hydrogen (Figure 5.14). This means that the equilibrium is favoured towards the conformer with the methyl group equatorial.

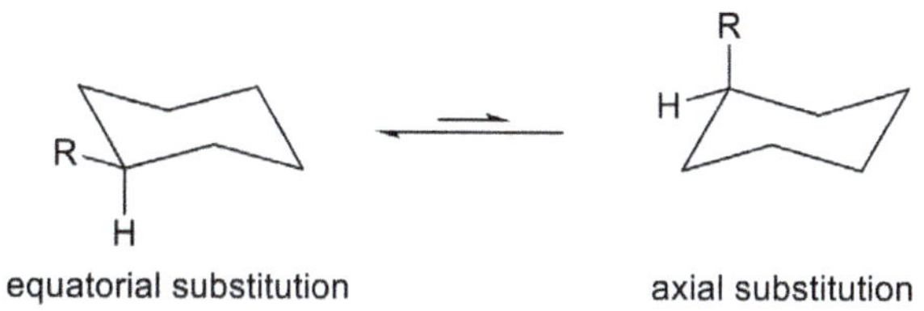

R	% Equatorial
H	50
Me	96
i-Pr	98
t-Bu	>99

equatorial substitution   axial substitution

Figure 5.14 Equilibrium distribution of various substituted cyclohexanes at room temperature.

When we increase the size of the substituent, the 1,3-diaxial interactions are even greater when they are axial. In fact, a tertiary-butyl group almost exclusively adopts the equatorial conformation, because it is so large.

When we have a cyclohexane with more than one substituent then the lowest-energy conformation will still be the chair with the minimal 1,3-diaxial interactions, but this will depend on both the relative orientation and size of the two groups. Again, we'll start with chairs containing methyl groups.

> Here we are just looking at the effect of the size of the substituent. Things get more complicated when you also consider electronic factors which you will see in Section 5.5.2.

If we consider *cis*-1,2-dimethylcyclohexane (Figure 5.15), it can exist as a chair conformation with one axial group and one equatorial group (note how both methyl groups are on the same face, as the stereochemistry dictates).

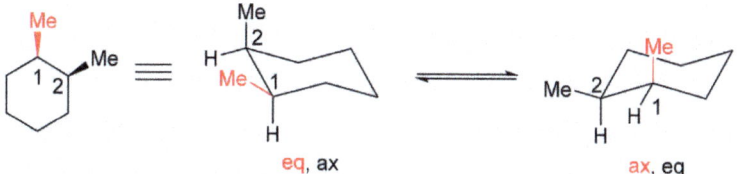

eq, ax   ax, eq

Figure 5.15 Conformations of 1,2-*cis*-dimethylcyclohexane.

When the first chair undergoes a ring flip, the equatorial methyl becomes axial (at carbon 1) and the axial methyl becomes equatorial (at carbon 2). Overall, both chairs have one axial and one equatorial methyl group, meaning the 1,3-diaxial interactions are the same. Hence, both chairs have the same energy and exist in equal amounts.

If we change the stereochemistry of one methyl group so that it is hashed (pointing away), one chair conformation can exist with both methyl groups being equatorial and the other with both groups axial (Figure 5.16). The both-axial chair will be much higher in energy than the both-equatorial chair so the equilibrium will favour the both-equatorial conformation.

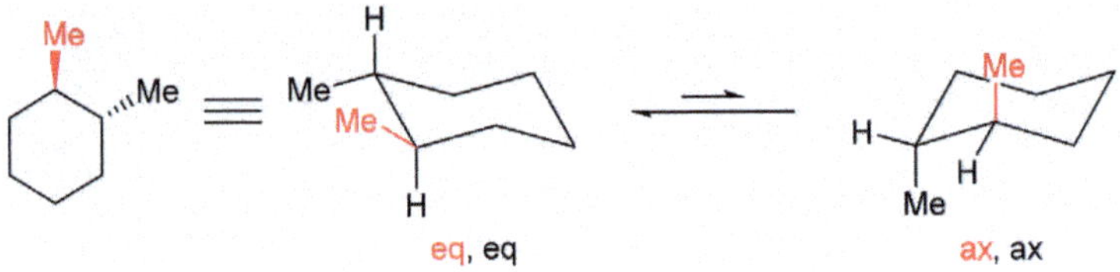

eq, eq   ax, ax

Figure 5.16   Conformations of 1,2-*trans*-dimethylcyclohexane.

What happens if we have different groups around the ring? The same principle still applies – the chair with minimal 1,3-diaxial interactions will be favoured. For example, if we consider the conformations of the common painkiller tramadol (Figure 5.17), the right-hand conformer is favoured for two reasons. Firstly, it has two of the three groups equatorial and secondly, they are the largest groups.

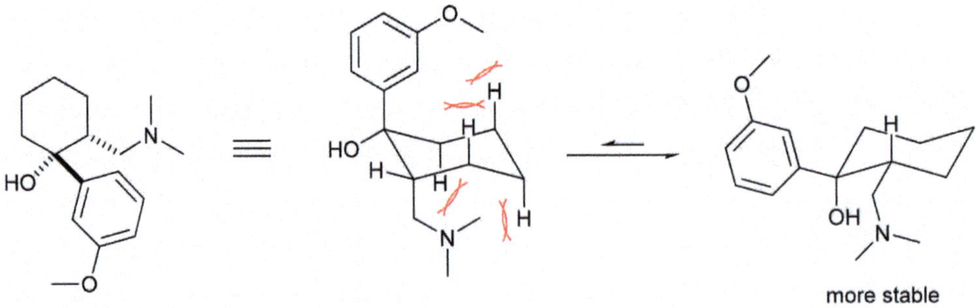

(±)-Tramadol

more stable

Figure 5.17   Conformations of tramadol.

Let's just check that you understand the difference between axial and equatorial substituents by identifying which of the substituted cyclohexanes in Figure 5.18 have functional groups in axial and equatorial positions.

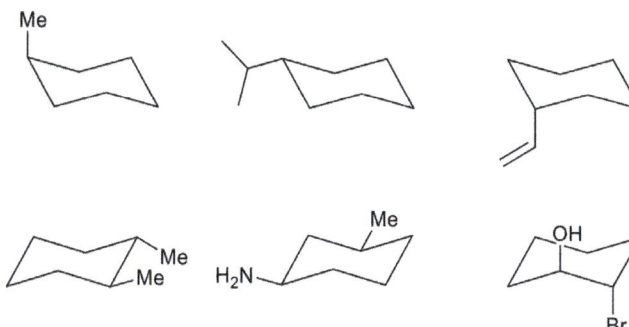

Figure 5.18   Substituted cyclohexanes.

The answer to the problem is given in Figure 5.19, where the three cyclohexanes with axial substituents are grouped together, and the three cyclohexanes with equatorial substituents are grouped together.

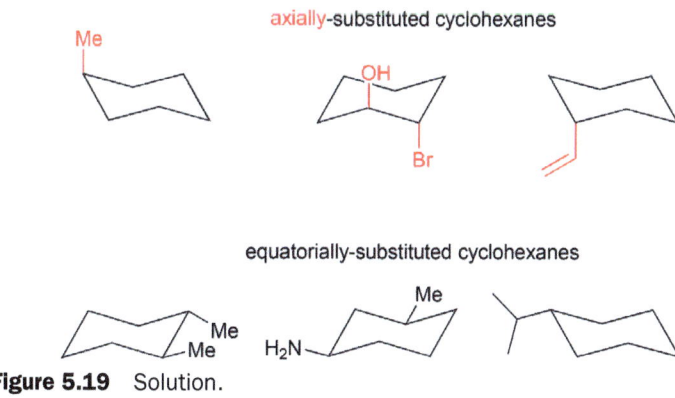

Figure 5.19   Solution.

What the axially substituted cyclohexanes all have in common is that the groups are either pointing above or below the ring – that's how we can identify them as being axial. Similarly, the equatorially substituted rings have their groups pointing to the side of the ring, so that's how we determine that they are equatorial.

5.2 Conformation of Rings Smaller Than Cyclohexane

With decreasing ring size, we find that rings experience more angular strain.

Cyclopentanes and other five-membered rings adopt envelope conformations where two adjacent carbons are in an eclipsed conformation (Figure 5.20). The strain for those two carbons can be relieved by ring flipping, but this results in another two carbons experiencing eclipsing interactions.

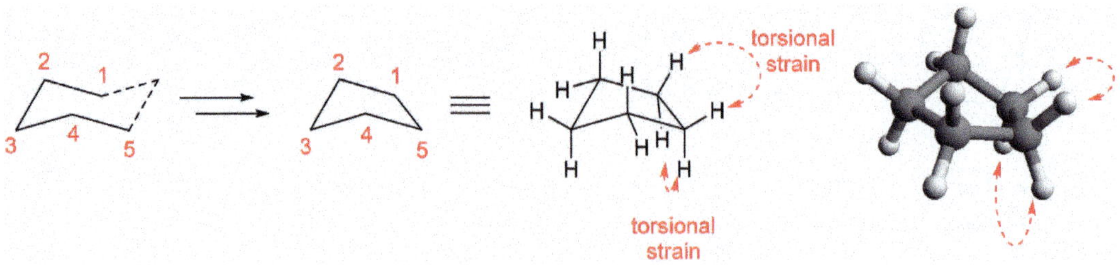

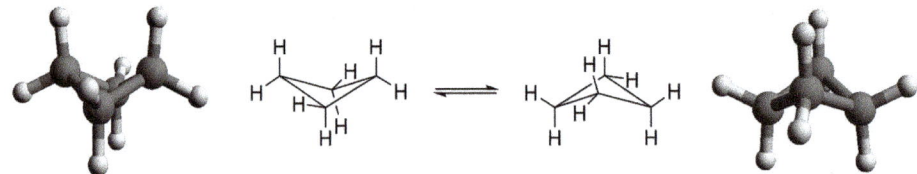

Figure 5.20   Conformations of cyclopentane.

Cyclobutanes (Figure 5.21) adopt a butterfly conformation and are the smallest rings that can invert. Ring flipping helps to relieve torsional strain of the axial hydrogens.

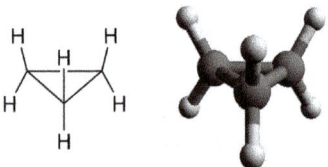

Figure 5.21   Conformations of cyclobutane.

Cyclopropanes (Figure 5.22) can only adopt a single conformation because they are triangular. The 60° bond angle between the carbons in the ring provides high angular strain. The hydrogens are in a fixed position and experience torsional strain.

Figure 5.22   Conformations of cyclopropane.

The following example is to check you've got the differences between the conformations of various ring sizes.

Please arrange the hydrocarbon rings shown in Figure 5.23 in order of stability and conformational flexibility.

■ Checkpoint

You should now understand the conformational analysis of six-membered rings and of other ring sizes.

 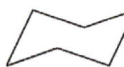 

Figure 5.23   Hydrocarbon rings.

That's right. As you decrease in ring size, the rings become more strained (Figure 5.24).

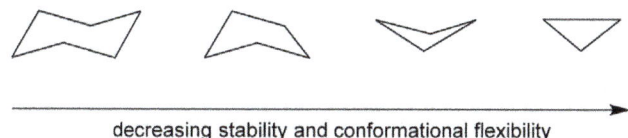

decreasing stability and conformational flexibility

Figure 5.24   Solution.

5.3 Rings Containing sp² Hybridised Carbons

The conformational effects we have discussed so far apply not only to fully saturated rings (where all carbons are sp³ hybridised), but also when one or more of the carbons contain sp² hybridised carbons.

5.3.1 Double Bonds Outside the Ring Are Neither Axial Nor Equatorial

If a cyclohexane is substituted with double bonds outside the ring (*e.g.* an alkene or ketone) the configuration of this group is neither axial nor equatorial, owing to the sp² hybridisation of the carbon. Overall, the ring will still have a chair conformation but the sp² hybridised carbon will adopt a trigonal planar shape. This results in these functional groups having a configuration that is in between where are an axial or equatorial substituent would be (Figure 5.25).

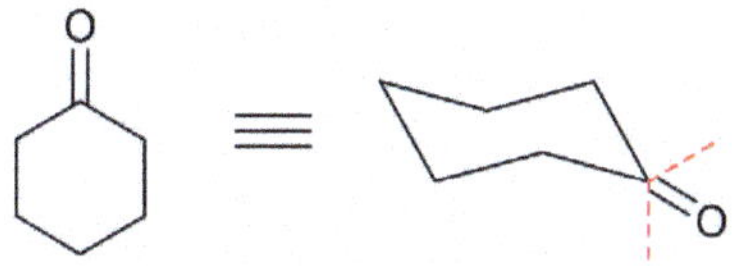

Figure 5.25   Conformation of cyclohexanone.

5.3.2 Double Bonds Inside the Ring Have a Half-chair Conformation

When there are double bonds inside the ring, *i.e.* a cyclohexene, there are two sp² hybridised, trigonal planar carbons, which results in one half of the ring being planar (Figure 5.26). We have already seen this conformation before – the half-chair.

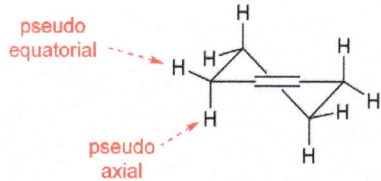

pseudo equatorial

pseudo axial

Figure 5.26   Conformation of cyclohexene.

The substituents on a half-chair conformation adjacent to the double bond are in similar positions to axial and equatorial groups but ring strain leads to them having slightly different bond angles. We call these groups **pseudoaxial** and **pseudoequatorial**.

A half-chair can ring flip (Figure 5.27) but only the sp^3 hybridised section (*i.e.* the chair part) does this. Just like a cyclohexane, pseudoaxial groups become pseudoequatorial and *vice versa*. Therefore, half-chairs will favour a conformation where the largest groups will be pseudoequatorial.

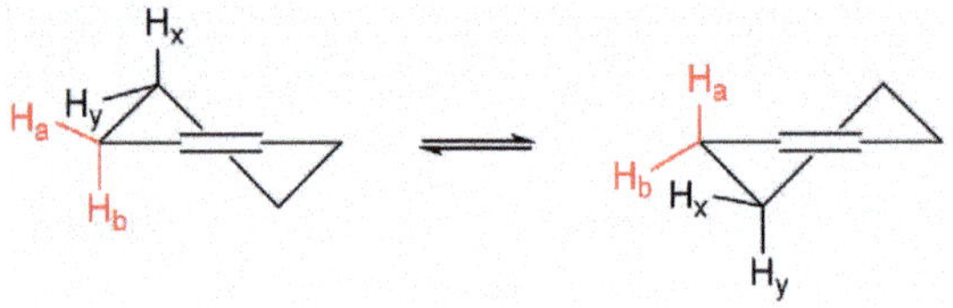

Figure 5.27   How substituents interconvert following a cyclohexene ring flip.

5.4 Fused Rings

Steroids are naturally occurring molecules that perform a variety of signalling roles in the body. Even though different steroids may have completely different biological functions, what links this family of compounds together is the fused ring system common to all steroids (Figure 5.28). In this section, we will introduce the features that fused ring systems have.

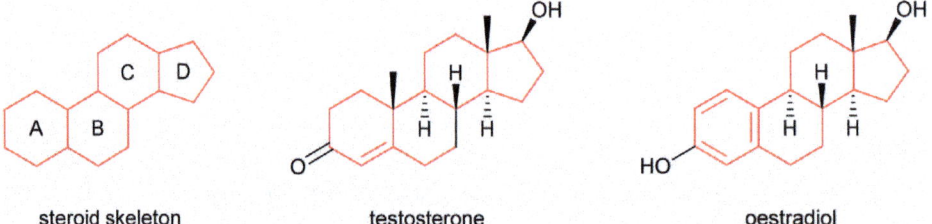

steroid skeleton   testosterone   oestradiol

Figure 5.28   Steroid structures.

5.4.1 Decalins

Two cyclohexane rings joined together are called a **decalin**, so called because it is made up of ten carbons. There are two types of decalin, based on the stereochemistry where the two rings join.

The first type of decalin is a *trans*-decalin (Figure 5.29) where the fused rings have *trans* stereochemistry. A *trans*-decalin exists as one conformation with the two rings adopting chair conformations, as a result of the second ring being attached to the first ring in equatorial positions.

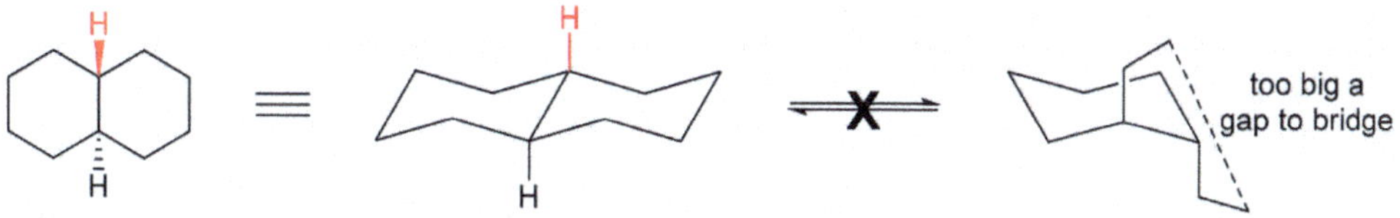

trans-decalin

Figure 5.29   Conformations of *trans*-decalins.

If the decalin were to flip, the second ring would be attached to the first in a *trans*-diaxial conformation. These carbons end up so far apart that the rest of the second ring cannot form.

The second type of decalin is called a *cis*-decalin. Here the second ring is attached on the same face. In *cis*-decalins both rings have chair conformations (Figure 5.30), which are joined in one axial and one equatorial position. A ring-flipped *cis*-decalin will have the same conformation, so the two conformers are in equilibrium with one another. The axial substituent means that *cis*-decalins will be higher in energy than *trans*-decalins, owing to 1,3-diaxial interactions.

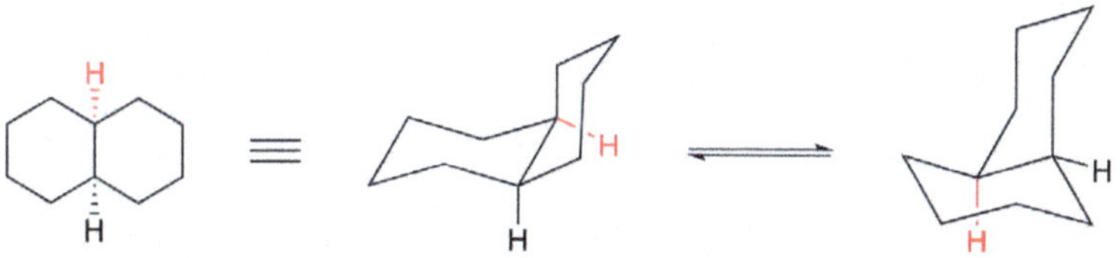

> Remember *cis* and *trans* here refer to the relative positions of the rings across the ring junction.

cis-decalin

Figure 5.30   Conformations of *cis*-decalins.

5.4.2   Fused Three- and Six-membered Rings

Epoxides (Figure 5.31), haloniums and aziridines fused onto a six-membered ring are useful intermediates in synthesis. For example, a fused aziridine is used as an intermediate in the synthesis of the antiviral drug oseltamivir (Tamiflu®), as shown in Figure 5.32.

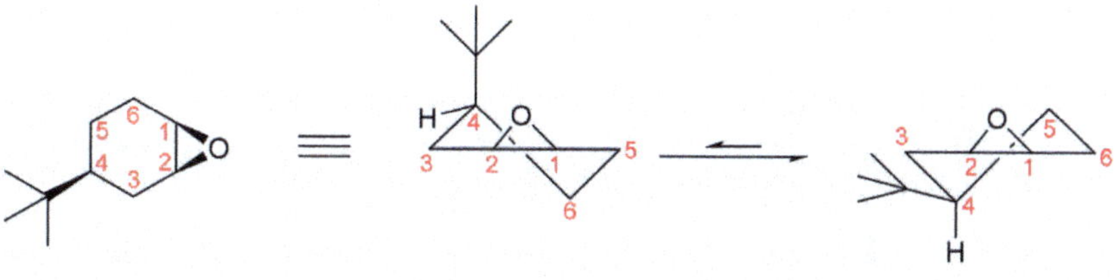

> It's sometimes easier to look at the relative positions of the hydrogen atoms. In a *trans*-decalin they will also be *trans*, in a *cis*-decalin they are *cis* to each other.

more stable

Figure 5.31   Conformations of fused epoxides.

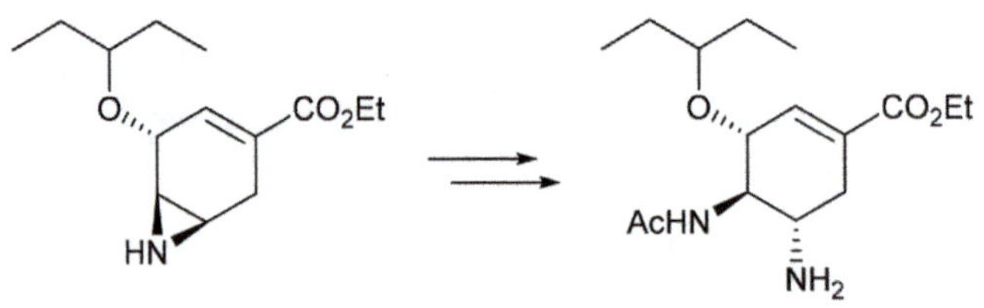

fused aziridine   oseltamivir (Tamiflu®)

Figure 5.32   Structure of oseltamivir and its precursor.

The highly strained three-membered ring places strain on the cyclohexane, so it has a half-chair conformation. Like cyclohexene, the ring can flip and the most stable conformer has the largest groups equatorial.

The orientation of the three-membered ring should also be considered. It will point up or down depending on its stereochemistry (Figure 5.33).

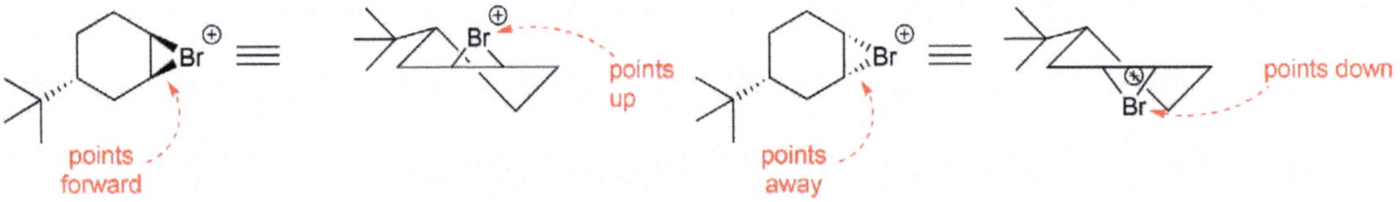

Figure 5.33   Correct orientation of 3-membered ring based on its stereochemistry.

Let's just check you understand how substituents and fused rings should be oriented in their half-chair conformations.

Two representations of the fused aziridine are shown in Figure 5.34. Which of these is a correct representation of the higher energy conformation of the aziridine shown?

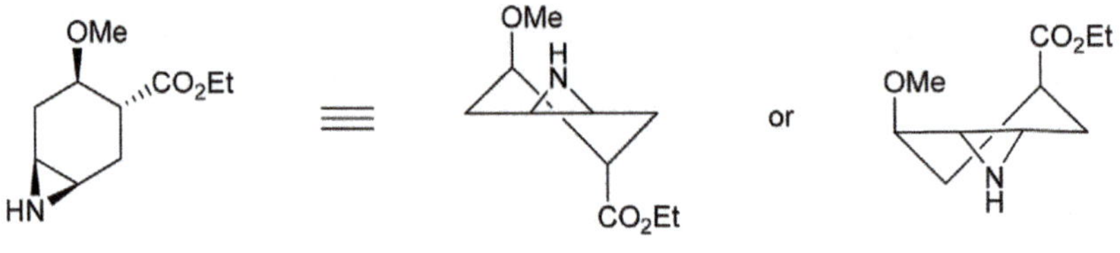

Conformation **A**   Conformation **B**

Figure 5.34   Which is the badly drawn and which is the correctly drawn representation?

The first stage in determining which is the correct conformation is to label the carbons of the 2D structure and note the orientation of the aziridine ring and substituents.

If you look at the 2D structure, you will see how the aziridine ring is pointing towards you and you will see the stereochemistry of the ethyl ester. The numbering of the ring shows where both the methyl ether and ethyl ester should be relative to the aziridine (Figure 5.35).

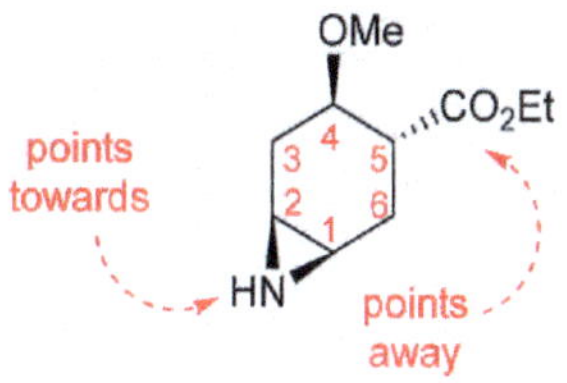

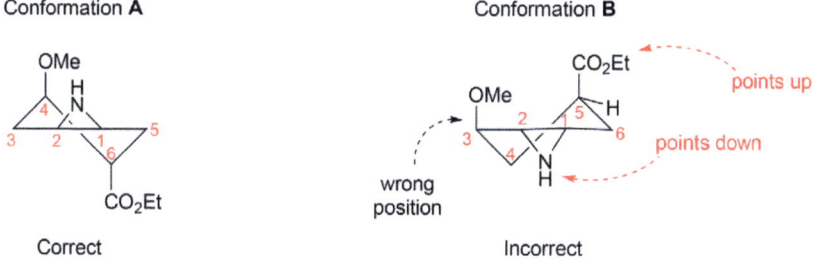

Figure 5.35   Labelling the carbons.

Looking at Figure 5.36, numbering the carbons shows that Conformation A has both substituents in the correct positions whereas Conformation B has the methyl ether on the wrong carbon (carbon 3). Conformation A has the aziridine pointing up so it matches the stereochemistry of the 2D structure, whereas Conformation B has the aziridine pointing in the opposite direction. Finally, Conformation B has switched the stereochemistry of the ethyl ester. In the 2D structure, it is pointing away so it should be pointing down in the 3D representation.

Figure 5.36   Both representations.

5.5   Chair Conformations in Nature: Sugars

At the beginning of this chapter, we explained how sugars perform a variety of biological functions.

5.5.1   Sugars Can Form Five and Six-membered Rings

We showed sugars as six-membered rings, but they can also form five-membered rings. Sugars can interconvert between these forms *via* a linear isomer based on carbonyl chemistry. In the linear form, the sugar has a terminal aldehyde group. One of the alcohols of the linear sugars can react with the aldehyde to form a hemi-acetal. Whether a five- or six-membered ring forms depends on which alcohol reacts. A sugar that is a five-membered ring is called a **furanose** sugar, and a sugar that is a six-membered ring is called a **pyranose** sugar (Figure 5.37). As we now know, the six-membered ring will be more stable in most cases because there is less angle strain in the ring and less torsional and steric strain between the substituents.

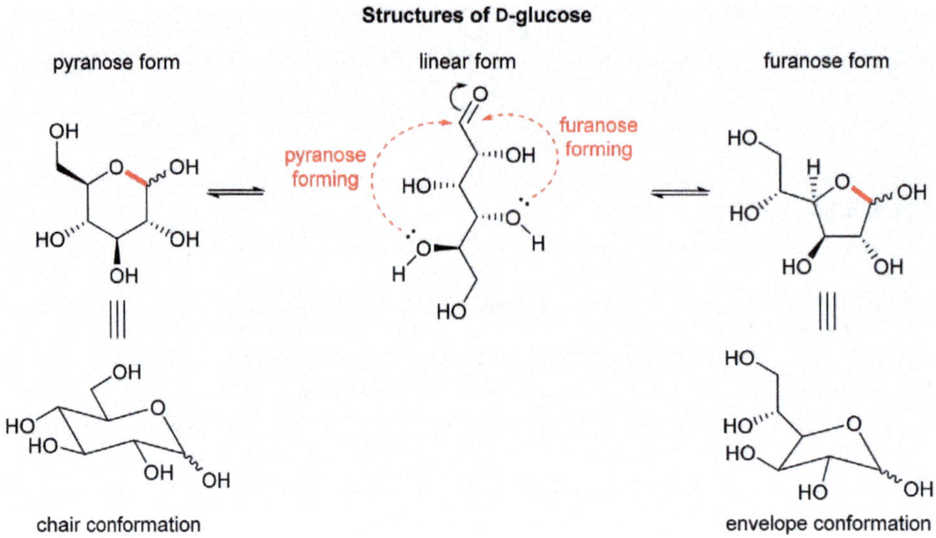

Figure 5.37   Pyranose and furanose form of D-glucose.

5.5.2   *The Anomeric Effect*

The carbon with the hemi-acetal **functionality** is called the anomeric carbon. Here, the OH that forms from the cyclisation can exist in the axial or equatorial position (Figure 5.38).

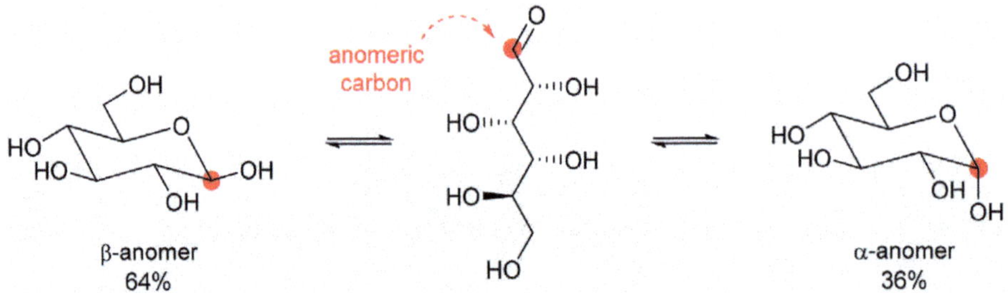

Figure 5.38   α and β anomers of D-glucose.

For glucose, the equatorial form is favoured over the axial form but the distribution towards the equatorial form isn't as strongly favoured as you'd expect. That is because there is an important orbital effect, called the **anomeric effect**, which stabilises the axial OH.

The two lone pairs of electrons from the oxygen in the ring are in axial and equatorial configurations (Figure 5.39). The axial lone pair can stabilise the axial OH group by feeding its electrons into the empty C–O σ* orbital. When the OH is equatorial, the orbitals do not line up together to allow the anomeric effect to occur. For those of you who like frontier molecular orbitals, the diagram on the right of Figure 5.39 may help.

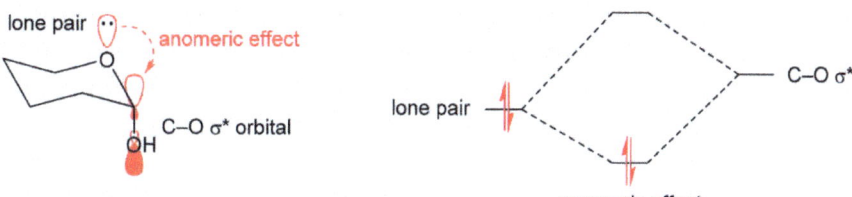

Figure 5.39 Frontier molecular orbitals involved in the anomeric effect.

---

❶ Key Learning Point

The anomeric effect may lead to a preference for the axial over the equatorial position, depending on the molecule you are studying.

5.5.3 The Difference Between Sugars Is Clearer In 3D

Glucose, galactose and mannose are diastereoisomers of each other. If we view them in chair form (Figure 5.40), we can see that glucose has all of its substituents in equatorial positions. Galactose and mannose each differ from glucose in the stereochemistry of

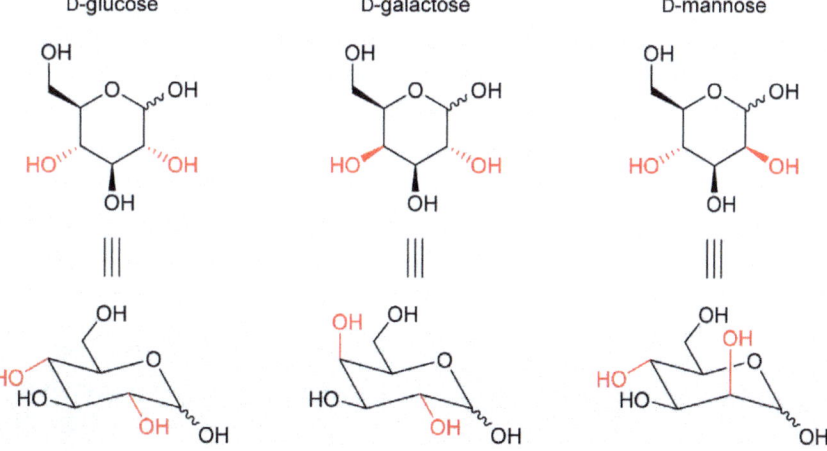

Figure 5.40 Chair conformations of some common sugars.

one alcohol, meaning it has to adopt an axial conformation. These small changes between the sugars mean that they are recognised differently by enzymes; this allows them to carry out different functions.

Let's just check you understand the steric and electronic effects of substituents on sugars.

Given that the acetoxy group has a significantly stronger anomeric effect than a hydroxyl group, which of the two conformations, in Figure 5.41, A or B, is more stable? Explain your answer using steric or electronic effects (or both).

Conformation **A**   Conformation **B**

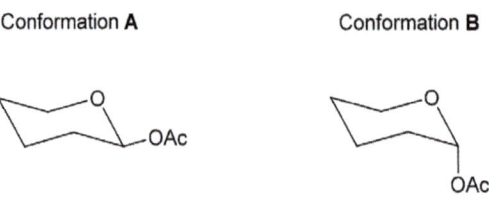

Figure 5.41   Which conformation is more stable?

Without question, practice makes perfect when it comes to drawing chairs so don't worry if your first attempts are not great. If you keep practising them, they will improve.

Conformation B is the more stable conformer, owing to the anomeric effect. While sterics determine that the acetoxy group is preferably in the equatorial position, the anomeric affect (which is an electronic effect) is stronger and so the acetoxy group prefers to be axial, as given in Conformation B.

The anomeric effect stabilises the acetoxy group in the axial position by feeding a lone pair of electrons from the oxygen atom in the ring into the empty C–O σ* orbital (Figure 5.42).

lone pair   anomeric effect

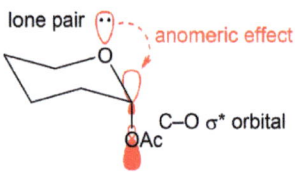

C–O σ* orbital

OAc

Figure 5.42   The more stable conformation, showing the anomeric effect.

▰ Checkpoint

You should now be able to draw conformations for cyclohexanes that are part of more complex structures.

5.6 Drawing Chair and Half-chair Conformations

So far, we have focused on the concepts of cyclic conformational analysis but for the rest of the chapter we will discuss how that knowledge is used in practice, starting with how to correctly draw chairs from 2D structures and *vice versa*.

5.6.1 Drawing Chairs

We'll start with how to draw the ring correctly. It is important that all the carbons are visible and that they have the correct bond angles, as a chair conformation shows.

You can think of a chair as being constructed from three pairs of parallel lines, so drawing one pair at a time will give you an accurate representation. Follow the steps shown in Figure 5.43 to draw a chair.

Step 1. Draw two parallel slanting lines that are level with each other.

Step 2. Add the second set of lines at an acute angle not going past the middle between the first two lines.

Step 3. Add the final set of lines to complete the ring.

Figure 5.43 Steps towards drawing a chair conformation.

What about drawing the flipped chair?

Start with drawing the first two parallel lines slanting in the opposite directions; then complete the rest of the chair in the same way (Figure 5.44).

Figure 5.44 How to draw the flipped chair.

5.6.2 Drawing Axial and Equatorial Substituents

Axial groups must point either up or down from the ring but how do you know which orientation each group should be on each carbon?

A simple rule is to look at whether the bonds going towards the carbon of interest point up or down. If they point up, the axial group points up, and if they point down then the group points down as well (Figure 5.45).

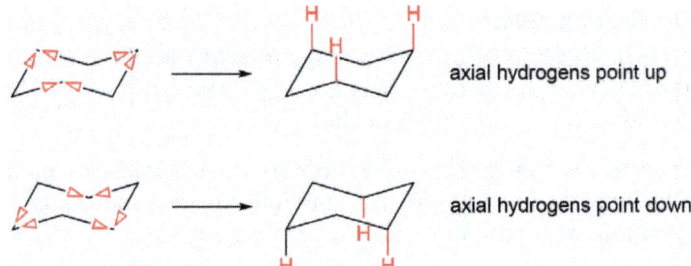

Figure 5.45   How to draw the axial hydrogens.

Equatorial groups should be drawn so that they are parallel with adjacent bonds. These are highlighted in red in Figure 5.46 so you can see which bonds the equatorial groups need to match.

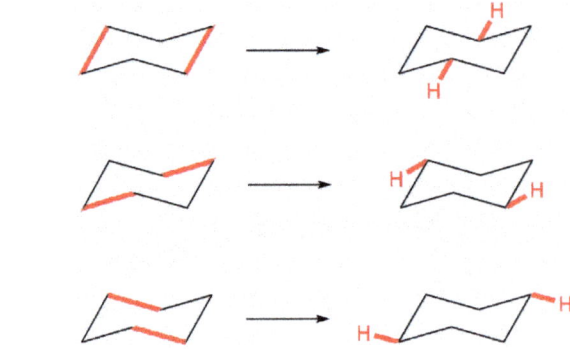

Figure 5.46   How to draw the equatorial hydrogens.

5.6.3   *Making Sure the Stereochemistry Is Correct*

One of the most common mistakes students make when drawing chairs is with the stereochemistry of the substituents and with their relative orientation around the ring. A mistake with either aspect means the chair represents a different molecule (Figure 5.47).

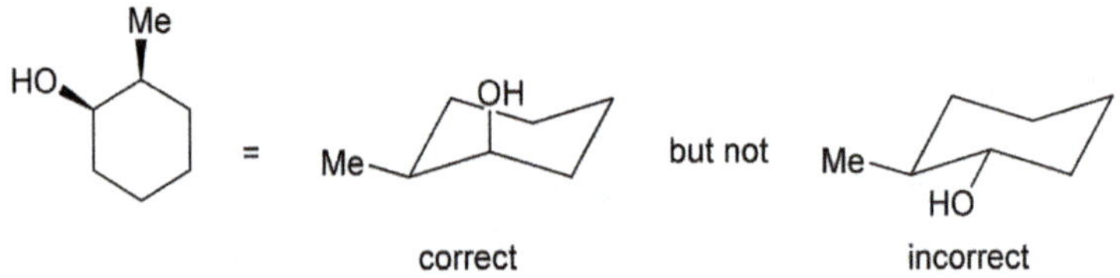

Figure 5.47   It is important to draw the correct stereochemistry of a chair conformation.

If the substituent in the 2D structure is wedged (*i.e.* pointing forward) then it should be on the top face in the chair. What can help to ensure that this is correct is to draw both bonds on the carbon so you can see clearly which is on the top face of the ring and which is on the bottom face.

If you want the group to be axial then there are positions in the ring where the methyl group will be both axial and on the top face (Figure 5.48).

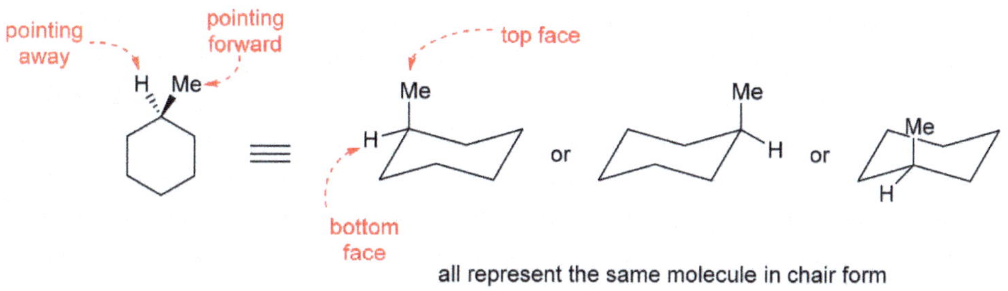

all represent the same molecule in chair form

Figure 5.48 How to draw the correct representation of axial substituents.

If you want the group to be equatorial, then the other three positions around the ring will be both equatorial and on the top face (Figure 5.49). If the substituent in the 2D structure is hashed (*i.e.* pointing away), then the same principle applies, except that the groups should be on the bottom face.

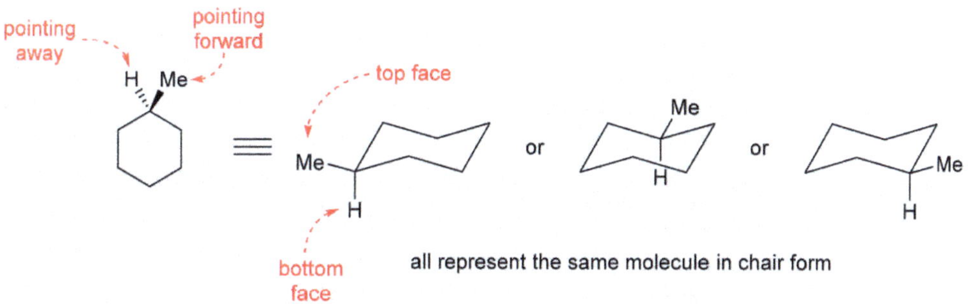

all represent the same molecule in chair form

Figure 5.49 How to draw the correct representation of equatorial substituents.

If you are drawing a chair with more than one substituent, it is still possible to make a mistake if you draw the stereochemistry (*i.e.* top and bottom faces) correctly. This can happen by drawing the groups in the wrong orientation around the ring. After drawing the first substituent the other groups must match the clockwise or anti-clockwise orientation of the groups around the ring (Figure 5.50).

Let's just check that you can accurately draw a chair conformation with all the substituents displaying the correct stereochemistry.

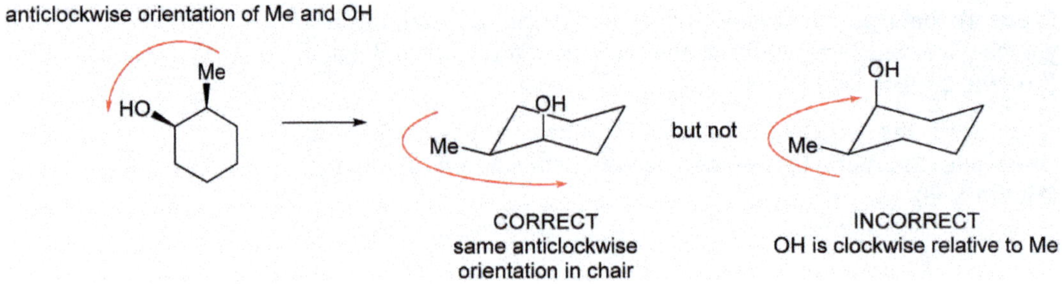

anticlockwise orientation of Me and OH

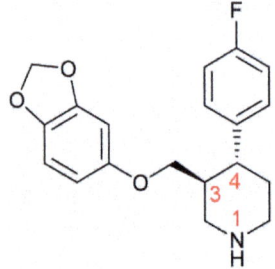

CORRECT
same anticlockwise
orientation in chair

but not

INCORRECT
OH is clockwise relative to Me

Figure 5.50   How to draw the correct orientation of substituents.

(−)-(3*S*,4*R*)-Paroxetine (Figure 5.51) is used to treat depression and is sold as a single enantiomer. Draw both conformations of paroxetine and identify which one is more stable.

Figure 5.51   (−)-(3*S*,4*R*)-Paroxetine.

First, we'll draw both chair forms. The bonds to the nitrogen need to be opposite in both chairs (*i.e.* upwards in one chair and downwards in the other) to correctly represent the ring flip (Figure 5.52). You could have chosen to view the ring from a different side. It doesn't matter.

Figure 5.52   Both chair forms.

We'll next add the fluorophenyl group. It needs to be on the opposite side of the ring to the nitrogen and on the bottom face for both chairs. This should result in the group being axial in one chair, and equatorial in the other (Figure 5.53).

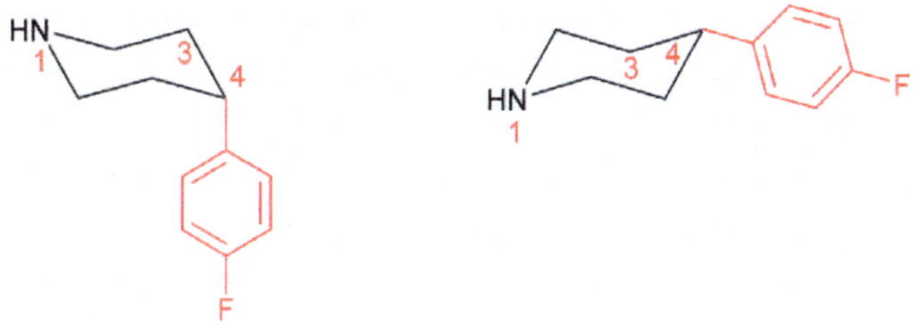

Figure 5.53 Both chair forms with the fluorophenyl group.

To have the correct stereochemistry of the ether substituent, it must be on the top face on the carbon next to the fluorinated aromatic on the anticlockwise side. As with the first group, one chair should have the group axial and the other equatorial.

Finally, we can show the equilibrium distribution to show how the all-equatorial chair is the lowest-energy and hence most prevalent conformer (Figure 5.54).

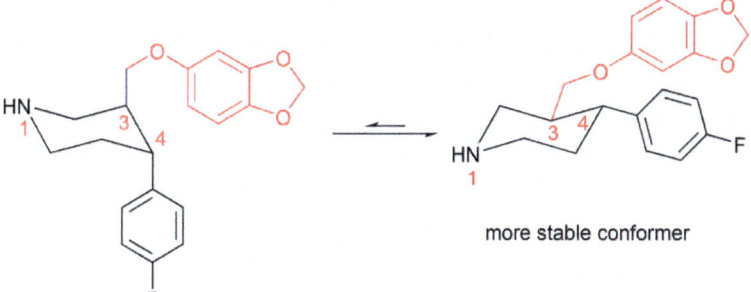

more stable conformer

Figure 5.54 The complete solution.

> Remember that each substituent will be axial in one chair conformation and equatorial in the other.

5.6.4 Drawing Half-chairs

To draw a half-chair conformation, first draw the planar portion of the ring at the front, then one rear carbon pointing up and the other pointing down (Figure 5.55). The flipped half-chair has the carbons pointing in opposite directions. The principles of drawing pseudo-axial or equatorial groups and having the correct stereochemistry is exactly the same as we have just discussed for chair conformations.

> **⚑ Checkpoint**
>
> You should now be able to accurately represent six-membered rings and their substituents in 3D.

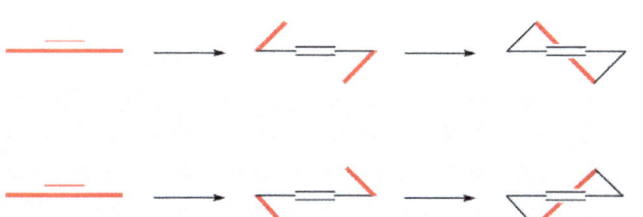

Figure 5.55 How to draw a half-chair conformation.

5.6.5   *From Chair to 2D Hashed and Wedged Structures*

If you want to draw the 2D structure from a given chair conformation, then draw the 2D structure as if you were viewing the chair from above (Figure 5.56). Whether a substituent is on the top or bottom face will tell you if the group should be hashed or wedged. Make sure the orientation is correct by following the connectivity of the groups in a clockwise direction.

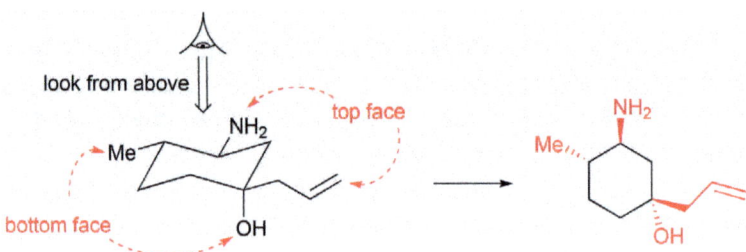

Figure 5.56   How to draw a 2D structure from a chair.

5.7   Structure Determination Using ^{1}H NMR

For chemists who are either isolating or synthesising cyclic molecules, it is important to be able to determine whether substituents are axial or equatorial. Fortunately, ^{1}H NMR spectroscopic analysis can be used to determine axial and equatorial configurations.

5.7.1   *Using the Karplus Relationship*

Owing to the Karplus relationship (which you came across in Chapter 4), we can distinguish between axial and equatorial protons. If you remember, the size of the coupling constant ^3J depends on the bond angle. If we look at the Newman projections of cyclohexane again, the bond angles become clearer.

A proton in the axial position will have a small coupling constant ^3J, of 3–4 Hz, to the equatorial hydrogen on the adjacent carbon, and a large coupling constant ^3J, of 10–12 Hz, to the adjacent axial hydrogen (Figure 5.57).

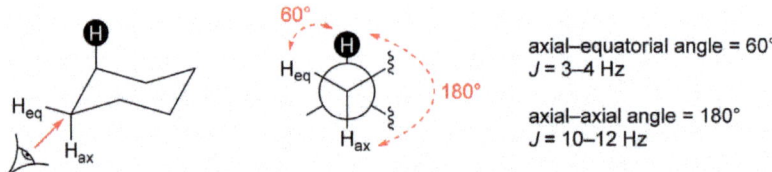

Figure 5.57   Bond angles and coupling constants of axial hydrogens to adjacent hydrogens.

A proton in the equatorial position will have a small coupling constant ^3J, of 3–4 Hz, to the adjacent equatorial hydrogen, and another small coupling constant ^3J, of 3–4 Hz, to the adjacent axial hydrogen (Figure 5.58).

> **❗ Key Learning Point**
>
> The largest coupling constants in cyclohexanes are usually due to two adjacent axial protons.

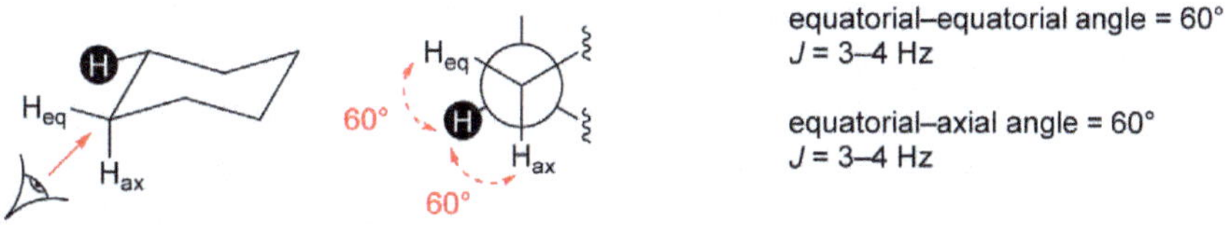

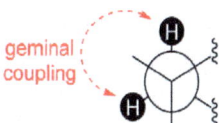

equatorial–equatorial angle = 60°
J = 3–4 Hz

equatorial–axial angle = 60°
J = 3–4 Hz

Figure 5.58 Bond angles and coupling constants of axial hydrogens to adjacent hydrogens.

5.7.2 Geminal Coupling

We've established that axial and equatorial hydrogens on the same carbon are not equivalent, so therefore it follows that they can couple to each other. We call this geminal coupling (Figure 5.59). The approximate size of geminal coupling is 12 Hz.

geminal
coupling

Figure 5.59 Geminal coupling.

5.7.3 Nuclear Overhauser Effect (nOe)

The nuclear Overhauser effect (nOe) is exploited in a ^{1}H NMR spectroscopic technique that shows which hydrogen atoms are close in space to one another (Figure 5.60). We know that 1,3-diaxial interactions occur because the protons are close in space, so nOe signals are seen between these protons.

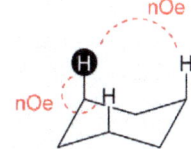

nOe

nOe

Figure 5.60 Nuclear Overhauser effect.

Let's just check that you can use a ¹H NMR spectrum to determine the chair conformation that is present.

(−)-(2S,6R)-Centrolobine (Figure 5.61) is a naturally occurring tetrahydropyran (a six-membered ring with an oxygen) that is used to treat the parasitic disease leishmaniasis.

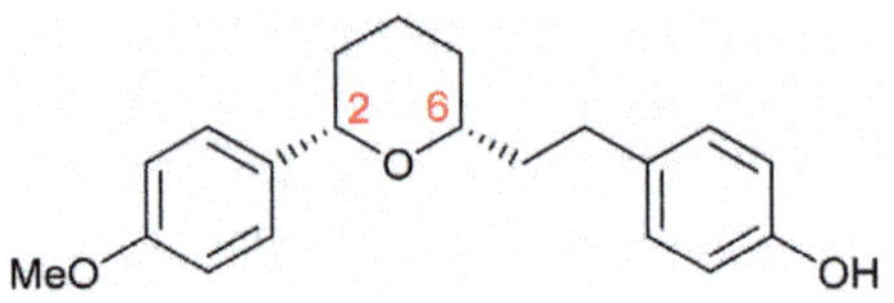

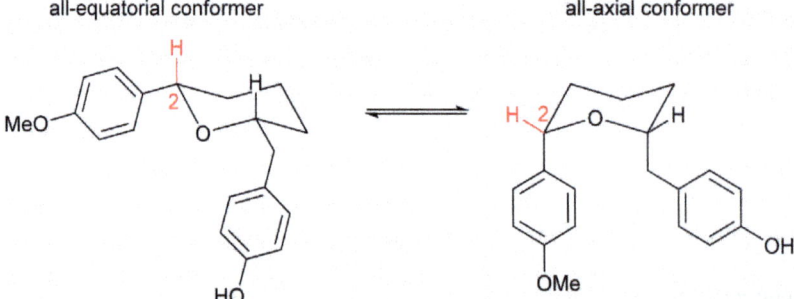

Figure 5.61   (−)-(2S,6R)-Centrolobine

We can draw two possible chair conformations of this molecule, the all-axial chair and the all-equatorial chair (Figure 5.62). We will show how analysis of the ¹H NMR peak at the 2-position tells us which conformer is present.

all-equatorial conformer   all-axial conformer

Figure 5.62   Two possible chair conformations.

You would expect the proton at the 2-position to appear as a doublet of doublets (dd), with an integration of one hydrogen and to be found between 4 and 5 ppm. For the all-equatorial conformer, you would expect one large and one small coupling constant, and for the all-axial conformer you would expect two small coupling constants.

The ¹H NMR spectrum has a doublet of doublets (dd) with integration of one hydrogen at 4.32 ppm (Figure 5.63). The two coupling constants are 1.3 and 11.1 Hz. The large coupling constant (11.1

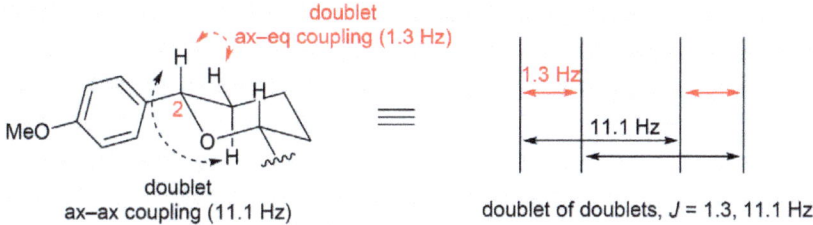

Figure 5.63   With the equatorial methoxyphenyl group.

Hz), in particular, confirms that the proton is axial, and therefore the methoxyphenyl group is equatorial.

This peak also has a nOe signal with the peak corresponding to the proton at the 6-position. This confirms that the proton at the 6-position is also axial and therefore (−)-centrolobine exists as the all-equatorial conformer (Figure 5.64).

> **◼ Checkpoint**
>
> You should now be able to apply conformational analysis in 1H NMR interpretation.

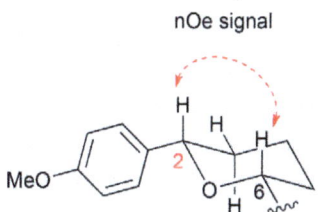

Figure 5.64   1,3-Diaxial protons give a nOe signal.

5.8 **Conclusion**

Well done! Representing ring structures in 3D is really tricky.

Hopefully you now feel confident in attempting conformational analysis of six-membered rings and of other ring sizes. You should also be able to represent six-membered rings and their substituents accurately in 3D and apply conformational analysis in ^{1}H NMR interpretation.

Finally, hold on to those terms 'axial' and 'equatorial'. You'll meet them again in Chapter 7.

Exercises

All these exercises test your ability to draw chair conformations correctly and then to answer short problems on conformational preferences, configuration or NMR spectroscopy. Each looks like a typical part of a stereochemistry exam question.

1. Draw both possible chair conformations for each of the following molecules. Highlight unfavourable interactions, and for each molecule identify which conformation is more stable.

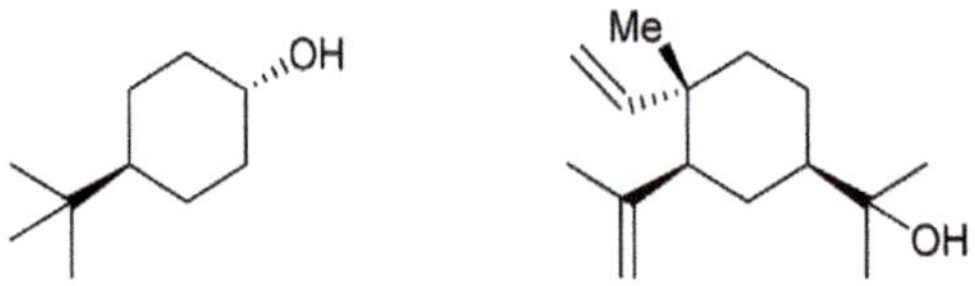

2. Draw half-chair conformations of (–)-limonene and identify which conformation is more stable.

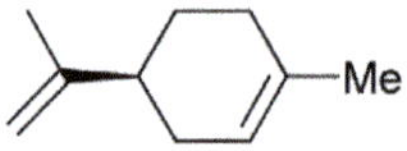

(–)-limonene

3. Assign the stereocentre of 2-methylpipridine as R or S (Hint: Draw the chair in its 2D format first).

4. Predict the multiplicity and coupling constants you would expect to see for both H_a and H_b in the following molecules.

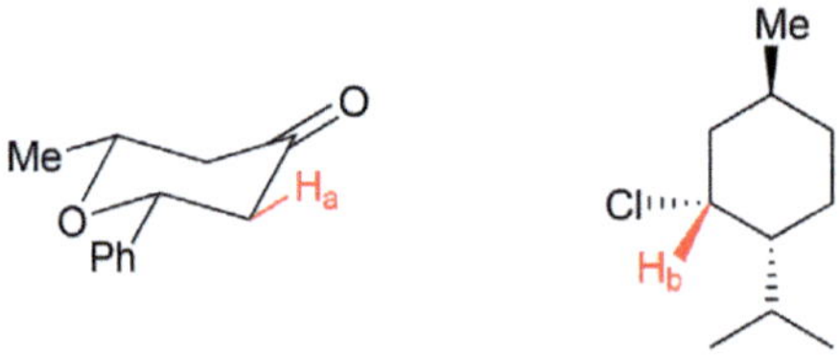

Answers

1. For the first cyclohexane, we can start by establishing that the tertiary-butyl group will be on the top face, and the alcohol will be on the bottom face. The hydrogens on each of these carbons have been added for clarity. The two groups are opposite each other so they also need to be opposite in their chair conformations.

   If the tertiary-butyl group is drawn so that it is equatorial in the left-hand chair, the alcohol will also be equatorial. When the chair is flipped for the right-hand chair both the tertiary-butyl group and alcohol will be axial. The most stable chair will be the left-hand chair, with these two groups equatorial to minimise 1,3-diaxial interactions.

For the second cyclohexane, the rules of establishing the correct stereochemistry are the same as for the first cyclohexane. What is more challenging in this question is how to obtain the correct orientation of the groups. Numbering the cyclohexane shows how the groups are oriented in an anticlockwise direction.

If we start by drawing the methyl group on carbon 1 as equatorial, we can then add the substituent on carbon 2 anticlockwise to the methyl and axial, to draw the correct stereochemistry. Continuing to carbon 4 in an anticlockwise direction, the alcohol group should also be axial. To flip the chair, start with carbon 1 again, this time with the methyl group axial. Following the carbons of the other substituents in an anticlockwise direction provides both these groups in the equatorial position.

The right-hand chair is more stable as it has more substituents and larger substituents in the equatorial conformation.

2. If we start by numbering the carbons on the cyclohexene, we can see that the propene substituent is opposite the methyl group. Next we can draw both half-chair conformations and number the carbons in the same way so that we can identify which is carbon 4. Adding the propene group with the correct stereochemistry so that it is on the top face shows that the right-hand half-chair conformation is more stable.

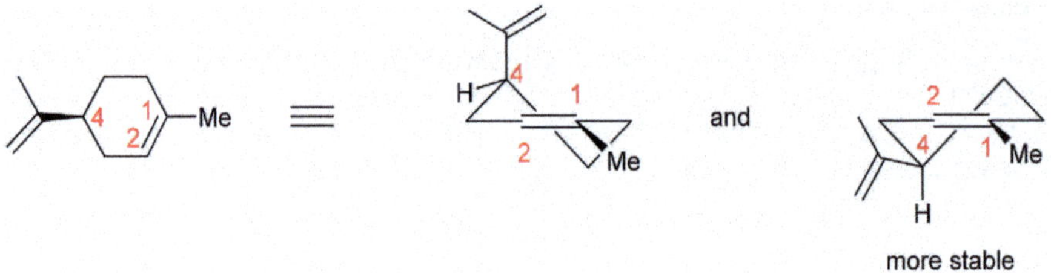

and

more stable

3. First redraw the chair in its 2D format. This makes it easier to then assign the priorities of the groups and their orientations using the method you have seen in Chapter 3. Priorities are assigned in order of the highest atomic number. We then look along the carbon to the bond with the lowest-priority group; this gives the first three groups in a clockwise order, meaning that the stereochemistry is R.

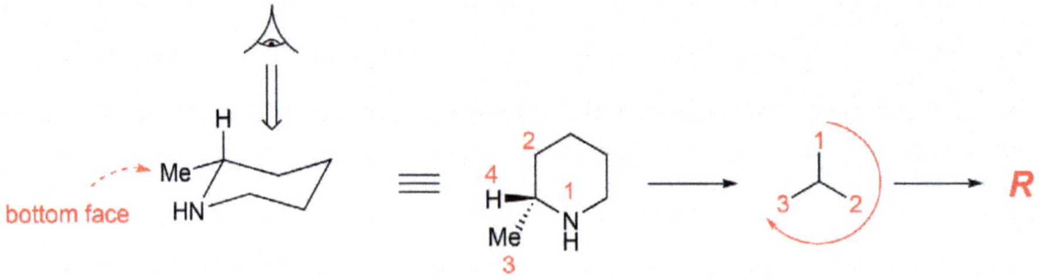

4. For the first molecule, H_a will be coupled to two protons – the axial proton on the same carbon and the axial proton on the carbon with the phenyl group; this will split the signal to give a doublet of doublets. With H_a being equatorial, this leads to an axial–equatorial coupling of about 4 Hz, and a diaxial coupling of about 12 Hz.

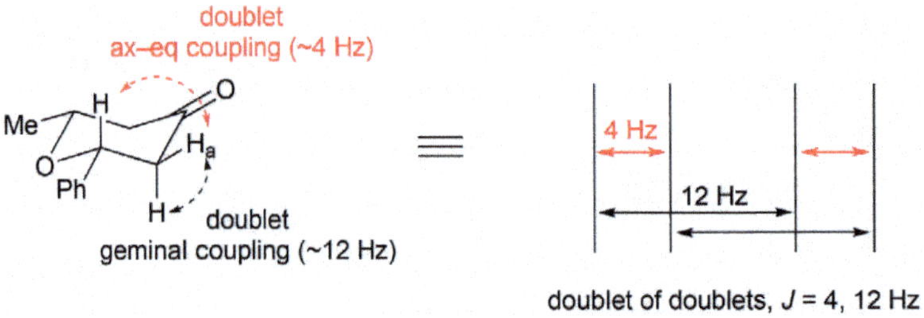

doublet
ax–eq coupling (~4 Hz)

doublet
geminal coupling (~12 Hz)

4 Hz

12 Hz

doublet of doublets, $J = 4$, 12 Hz

For the second molecule, H_b will couple to three different protons. There is an axial proton on either side, giving coupling constants of about 4 Hz and coupling to an equatorial proton, which will also be about 4 Hz. Because all three couplings are equal in size, the signal, which is actually a doublet of doublet of doublets, will appear as an apparent quartet.

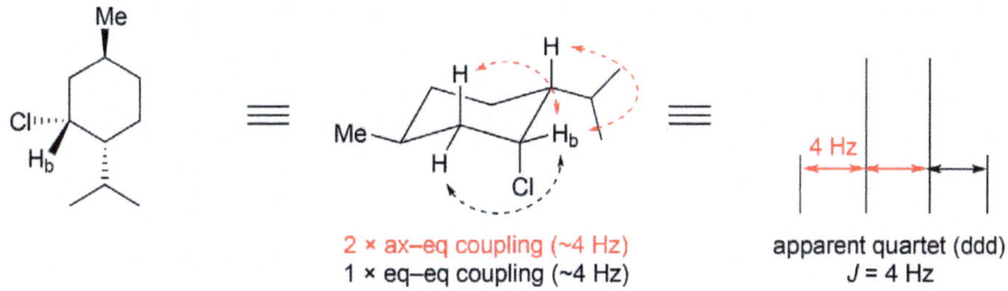

2 × ax–eq coupling (~4 Hz)
1 × eq–eq coupling (~4 Hz)

4 Hz

apparent quartet (ddd)
$J = 4$ Hz

By the End of This Chapter You Will:

☐ Understand how molecules *without* a stereogenic atom can be chiral.
☐ Be able to identify molecules with helical chirality (such as helicenes) and assign the configuration of helically chiral molecules using the *M* and *P* descriptors.
☐ Be able to define the term atropisomer and understand the reasons that lead to axial chirality in molecules.
☐ Understand what makes biaryl derivatives axially chiral and learn to assign the configuration of enantiomers using either the *R*/*S* system of descriptors or the *M* and *P* descriptors.
☐ Understand what makes allene derivatives axially chiral and learn to assign the configuration of enantiomers using either the *R*/*S* system of descriptors or the *M* and *P* descriptors.
☐ Appreciate the importance of axial chirality in organic synthesis and biology.

What You Will Get from This Chapter

We have two main tips for success in Chapter 6. Firstly, make sure you understand that helices are chiral. There are plenty of examples in the world around to help you explore this. Secondly, make sure you 'get' that allenes, hindered biaryl compounds and, more obviously, helicenes twist about an axis. You'll then be able to see the axial chirality that is the main new concept here. And look out for our insistence that it doesn't matter from which end you look at a helix when assigning its chirality.

CHAPTER

6

Chiral Molecules Without a Stereogenic Atom

6.1 Chirality Without Stereogenic Atoms

The structure of the naturally occurring antibiotic mycomycin (compound 6.1) is shown in Figure 6.1. It is isolated from the fungus *Nocardia acidophilus* in one enantiomeric form. How can this be possible, as there is no stereogenic centre? The **bidentate** phosphine ligand BINAP (compound 6.2) exhibits chirality. One enantiomer has been used as a ligand to make a chiral catalyst. Is this possible, as there is no stereogenic centre? The potential drug compound 6.3 (a glycine transporter inhibitor) exists as a pair of enantiomers that can be separated at room temperature. Both enantiomers have very different biological activities. Once again there is no stereogenic centre. Finally, the aromatic molecule [6]-helicene (compound 6.4) is chiral, a property that allows derivatives to be used in organic electronics. How can this be possible, as there is no stereogenic centre?

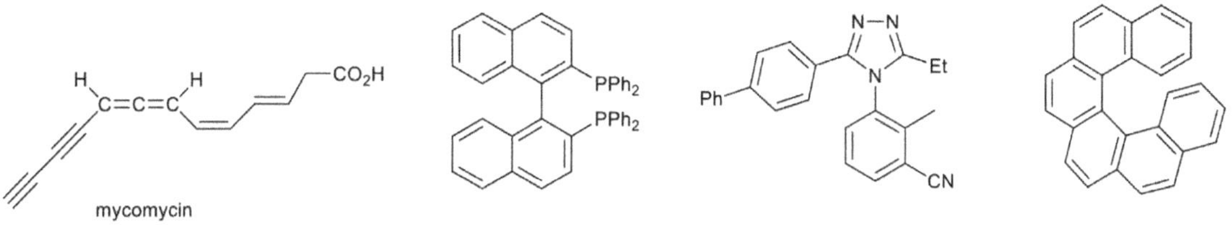

mycomycin **6.1** | BINAP **6.2** | **6.3** | [6]-helicene **6.4**

Figure 6.1 Some examples of molecules that are chiral.

In this book (Chapters 3–6), we have mainly concerned ourselves with chiral molecules that have stereogenic centres, such as (*S*)-lactic acid (compound 6.5), which is chiral, owing to its stereogenic carbon atom, highlighted in Figure 6.2. But, believe it or not, molecules (such as compounds 6.1 to 6.4) can be chiral even if they do not possess a stereogenic atom! The same rules apply as elsewhere in this book (Chapter 3), namely that if the mirror image of an object is not superimposable on the original then the object is chiral. You are already familiar with a number of everyday objects, such as screws and shells, that do not have a 'stereogenic centre' but do exhibit chirality (Figure 6.2). Instead of a stereogenic centre (or atom), these objects have a **stereogenic axis**, or '**axis of chirality**'. For a helix, this is the imaginary line running through the centre of the helix. Famously, DNA also forms helices.

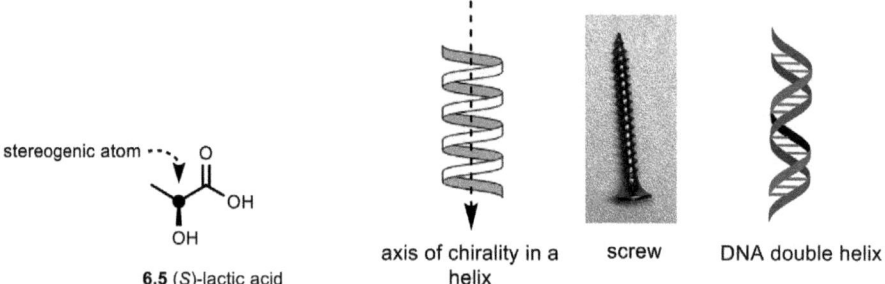

stereogenic atom

6.5 (*S*)-lactic acid

axis of chirality in a helix

screw

DNA double helix

Figure 6.2   (S)-lactic acid and some examples of natural helical structures exhibiting helical chirality.

This type of chirality is important to study and identify because molecules of this type are increasingly being used in asymmetric catalysis and smart materials but perhaps more significantly in drug discovery programmes, where testing individual enantiomers of molecules for biological activity is essential.

6.2 Visualising Helical Chirality

At first sight, it might not be obvious to see why helical systems are chiral. It is more easily visualised by considering the helical shells A and B, shown in Figure 6.3. These are mirror images of one another. They are also not superimposable on each other and so are chiral. If one traces the 'sense' of the spiral from the red dot in the centre, it can be seen that shell A contains a clockwise spiral (sometimes called a right-handed or dextral helix) while shell B has an anticlockwise one (a left-handed or sinistral helix). In Nature, most shells (over 90%) occur in the clockwise or dextral form. We officially classify shells by first looking down the helical axis (for shell C of Figure 6.3, this is the axis around which the spiral turns). Starting with the aspect of the shell closest to us we then determine whether the spiral follows a clockwise or anticlockwise direction (Figure 6.3). If you can find a snail or other spiral shell, you can verify this for yourself by looking from both directions. This helical axis is an example of an axis of chirality.

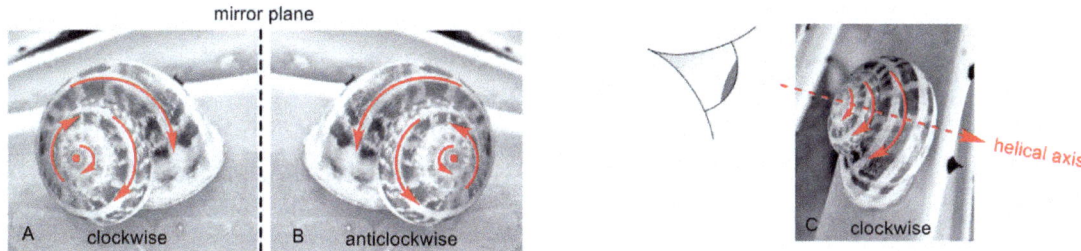

Figure 6.3   Chiral shells. How to visualise a helical axis.

6.3 Helicenes and Axial Chirality in Spiral Molecules

We have already seen that DNA exists as a double helix but one of the simplest classes of molecules to exhibit this form of helical chirality is that of the [*n*]-helicenes (*e.g.* compound 6.6 in Figure 6.4). These molecules have interesting chiroptical properties and have been used most recently as switches and sensors in organic electronics and materials chemistry, as well as in enantioselective catalysis.[1]

[*n*]-Helicenes are aromatic molecules that consist of a number (*n*) of fused benzene rings. [6]-Helicene (compound 6.6 in Figure 6.4) has been resolved into enantiomers. For this molecule, both rings A and F cannot occupy the same 3D space (owing to steric hindrance) and this leads to the molecule no longer being able to be planar, with one ring partially on top of the other (in this case, Ring A is shown above Ring F). This leads to the beginning of a helix (better visualised in the sideways view) and hence axial chirality (with the axis of chirality, or stereogenic axis, running through the centre of the spiral). If compound 6.6 were planar and consequently the rings A and F did point towards each other, the *atoms* highlighted in bold in compound 6.6 would have to occupy the same physical

It doesn't matter which way we look down the helical axis; if we start with the part of the shell closest to our eye, we always get the same answer.

▉ Checkpoint

You should now understand how molecules without a stereogenic atom can be chiral!

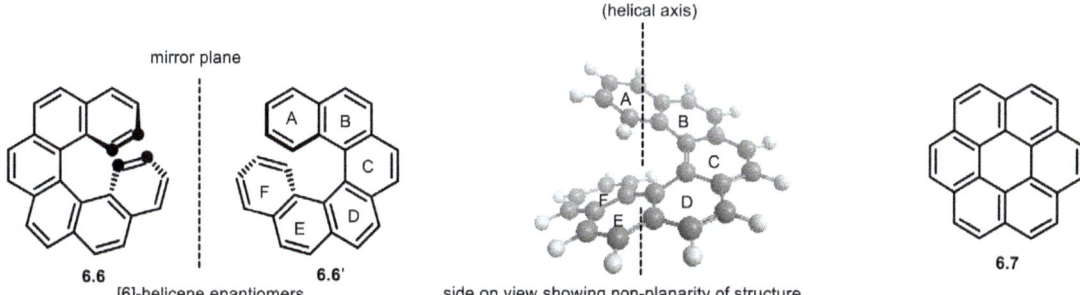

6.6   6.6'
[6]-helicene enantiomers

side on view showing non-planarity of structure

6.7

Figure 6.4   [6]-Helicene (compound 6.6) exhibits helical chirality around the chiral axis.

space. This would only be possible if the molecule fused to make compound 6.7. This molecule exists but it is a *different* molecule from compound 6.6.

> It is difficult to make a model of [6]-helicene 6.6 as most model kits will not have enough sp² carbon atoms. You could combine two or three kits with friends.

> **⚠ Key Learning Point**
>
> Helices are chiral.

How do we assign enantiomeric configuration descriptors to helicenes such as 6.6? Instead of using *R* and *S* nomenclature to assign enantiomeric configurations (see Section 3.2) we use the *P*/*M* system.

6.4 Assigning *P* and *M* Descriptors

Step 1. Firstly, we identify the stereogenic axis. For [6]-helicene, it is represented by the black dot in Figure 6.5 or the dashed line in the sideways viewpoint in Figure 6.4.

Step 2. Then we look down the axis – we can choose to look from *either* direction – *it doesn't matter*. In this example, we are *choosing* to look down from the top.

> **◄ Checkpoint**
>
> You should now be able to identify molecules with helical chirality (such as helicenes) and assign the configuration of helically chiral molecules using the *M* and *P* descriptors.

Step 3. We then draw a curved arrow from the part of the molecule *closest to us* (Ring A) to that furthest away (Ring F).

Step 4. If this curve is bending in a clockwise direction, we assign the *P* descriptor (*P* stands for '*plus*'). If the curve is bending in an anticlockwise direction, we assign the *M* descriptor (*M* stands for '*minus*'). In this case, the structure shown in Figure 6.5 is assigned as the *P* enantiomer.

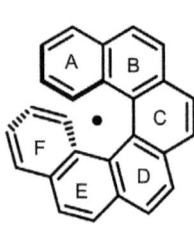

6.6

(i) Identify stereogenic axis.

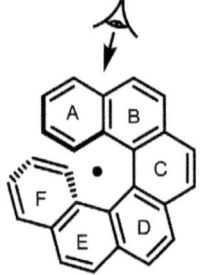

(ii) Look from above (ring A is closest to eye).

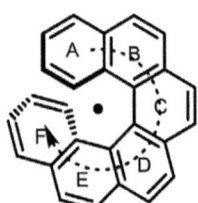

(iii) Draw curved arrow from closest ring (A) to furthest ring (F).

(iv) Clockwise arrow = *P* configuration.

Figure 6.5 Assigning *P* or *M* configuration to a molecule that is a helix.

6.5 Rotation Around Single Bonds (Revision)

In Chapter 4, you learnt that free rotation around the axis of the C–C single bond (σ bond) in ethane is rapid at room temperature and occurs at a frequency of $\approx 10^{11}$ s^{-1}. The energy barrier between the staggered and the eclipsed forms was measured at 12 kJ mol^{-1}, (R = H; the solid line in Figure 6.6).

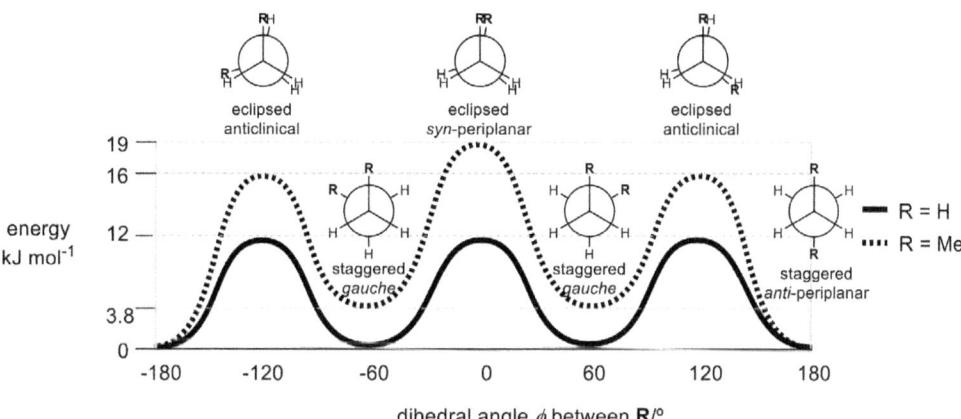

Figure 6.6   Revision of rotation barriers around the central C–C bond in ethane and butane.

---

❶ Key Learning Point

The IUPAC definition of atropisomers is 'a subclass of conformers that can be isolated as separate chemical species and arise from restricted rotation about a single bond', from the Greek "atropos", without turn.

Increasing the size of the substituents normally increases the energy barrier between conformations due to steric factors and this, in turn, slows the rate of rotation, hence rotation around the central C–C bond in butane was measured at 2×10^9 s^{-1} at room temperature (R = Me; the dotted line in Figure 6.6). If the energy barrier between individual conformers is high enough (>100 kJ mol^{-1} at room temperature), it is theoretically possible to separate different conformers arising from rotation around single bonds at room temperature; these conformers are called **atropisomers** and we will see examples of them later.

6.6 Rotation Around the C–C Single Bond in Biphenyl[2]

Biphenyl (Figure 6.7) consists of two aromatic rings joined by a single C–C bond (the carbon atoms of this C–C bond are highlighted as dots in Figure 6.7). The lowest-energy conformation is the 'twisted'

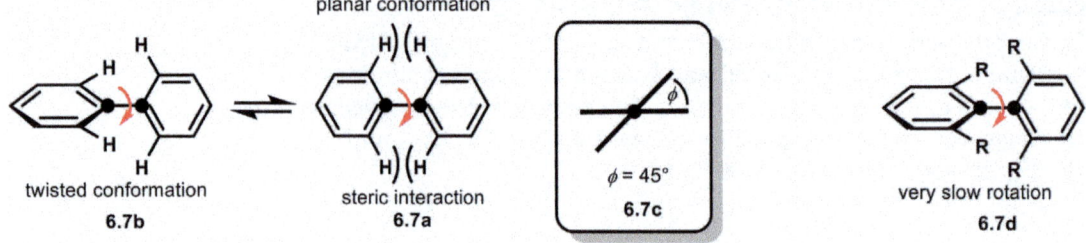

Figure 6.7   Conformation of biphenyl.

conformer 6.7b and not the planar conformer 6.7a, owing to steric repulsion between the four hydrogen atoms in the planar conformation. In the gas phase, the dihedral angle, ϕ (twist angle, Figure 6.7c), between the planes of both rings in the lowest-energy conformation has been measured to be about 45°. This is the angle that balances reducing steric clashes with maintaining some conjugation between the two rings. Owing to the steric clashes in the planar conformer (compound 6.7a) complete 360° rotation around the C–C bond is slower than in butane but is still fast enough that individual conformers are freely interconverting at room temperature.

If we replace the four *ortho* hydrogen atoms shown in Figure 6.7 (highlighted in 6.7a) with four larger *ortho* substituents (6.7d, R ≠ H), then the corresponding planar biaryl conformer can become so sterically hindered that the rate of rotation around the C–C bond is very slow, severely restricted or might not happen at all!

> 💬
> The carbon atoms highlighted by dots in compound 6.7d are sp^2 hybridised. Remember this for later.

> 💬
> Section 6.7 is to help you think about how stereogenic axes relate to stereogenic centres. Don't worry if you don't get it straight away!

6.7 A Thought Experiment (A Stereogenic Axis)[3]

What happens if we take a point in space and stretch it? We end up with a line, as shown in Figure 6.8. What has this to do with chirality? Let's undertake a little thought experiment to find out. If we take a chiral molecule with a stereogenic carbon atom, such as molecule A in Figure 6.9, it is clear there is a central stereogenic atom with four different substituents (R^1, R^2, R^3 and R^4) arranged at the vertices of a tetrahedron. The central carbon atom is analogous to the point in our example in Figure 6.8. We know that the mirror image of such a molecule is not superimposable and the system is chiral. Imagine if we could elongate and stretch this tetrahedron. Let's keep R^3 and R^4 in the same position in space and stretch the right-hand side to give molecule B. What has happened? We have an 'elongated distorted tetrahedron'; the four substituents R^1, R^2, R^3 and R^4 are still arranged at the vertices. We have 'stretched' the stereogenic centre (carbon atom) to give a line, the stereogenic axis.

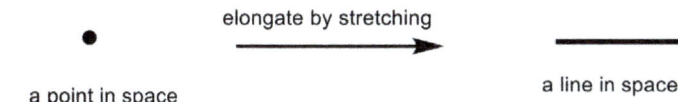

Figure 6.8   Stretching a point in space leads to a line.

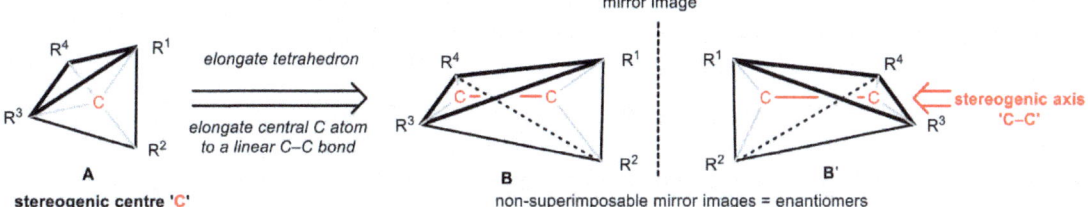

Figure 6.9   Stretching a chiral tetrahedron with a stereogenic centre leads to a distorted chiral tetrahedron with a chiral axis.

In chemistry, that will be any single bond, such as a C–C or C–N bond. Substituents R^1 and R^2 are attached to one end of the C–C bond and R^3 and R^4 to the other carbon of the bond.

It turns out that this distorted tetrahedron is still chiral. The mirror image, compound B′, is not superimposable on compound B. The enantiomers can theoretically be separated, *as long as there is no free rotation around the central C–C bond*. The new 'stretched' line in red (in this case the C–C bond) is called a stereogenic axis. More about this follows in the next section.

6.8 Axial Chirality in Substituted Biaryls[4]

In 1922, Christie and Kenner studied the chemistry of the biphenyl derivative shown in Figure 6.10 (6.8a). The molecule contains four *ortho* substituents and consequently, the planar conformer (6.8b) is not significantly populated at room temperature due to severe steric hindrance. Rotation around the C–C single bond

> Normally, in real-life examples of atropisomers, both carbons of the C–C single bond are sp^2 hybridised.

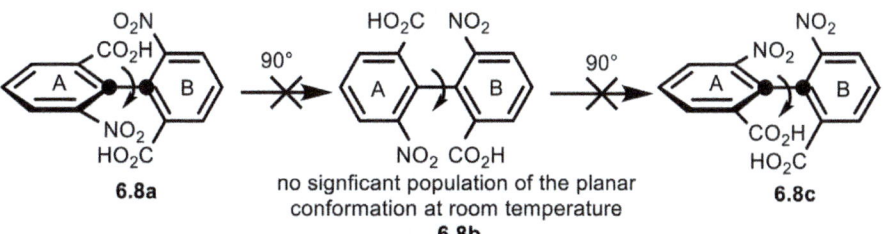

Figure 6.10   Atropisomers of biphenyl derivatives (6.8a and 6.8c) can be separated at room temperature.

linking both aromatic rings A and B is therefore very restricted and the energy barrier required to convert compound 6.8a to compound 6.8c (a rotation of 180°) is so high (>150 kJ mol⁻¹) that it is possible to separate the two conformers 6.8a and 6.8c. Remember our definition of atropisomers from Section 6.5? These conformers are atropisomers because they can be isolated as separate chemical species arising from restricted rotation around a single bond.[4]

Are the atropisomers 6.8a and 6.8c related in any other way? Are they the same or different? The two atropisomers 6.8a and 6.8c are not identical. You can tell this because both conformers 6.8a and 6.8c are drawn with the same orientation of ring B, but the orientations of substituents in Ring A are different in both molecules, (Figure 6.10). It is less straightforward to identify that 6.8a and 6.8c are, in fact, non-superimposable mirror images of each other and hence they are enantiomers.

The best way to convince yourself that compounds 6.8a and 6.8c are enantiomers is to make a model and compare them, but as with helicenes it is unlikely you will have enough sp² carbon atoms in your model kit, so you may have to borrow a friend's model kit.

However, you may be persuaded by Figure 6.11. It is relatively straightforward to draw the mirror image of compound 6.8a (Figure 6.11a, compound 6.8c'). We need to persuade you that compounds 6.8c' (Figure 6.11a) and 6.8c (Figure 6.10) are identical. If we look down the axis of the C–C single bond in 6.8c' from the right of the page (the axis is represented by the red line in Figure 6.11b) and rotate the whole compound 6.8c' by 90° in the direction of the black arrow (clockwise from our observation point), then we arrive at 6.8c. Hence, compounds 6.8c' and 6.8c are the same and we have proven that atropisomers 6.8a and 6.8c are enantiomers!

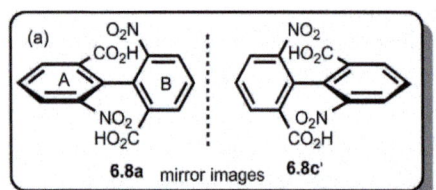

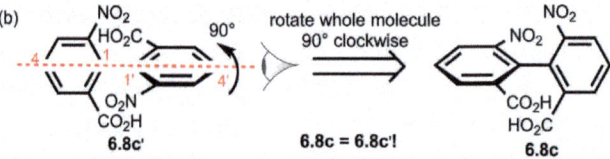

Figure 6.11   Enantiomers 6.8a and 6.8c'. Rotation shows that compound 6.8c is an enantiomer of compound 6.8a.

We can look at this another way by recognising that biaryl 6.8a can be viewed as a distorted chiral tetrahedron (Structure C in Figure 6.12). The C–C single bond with restricted rotation in compound

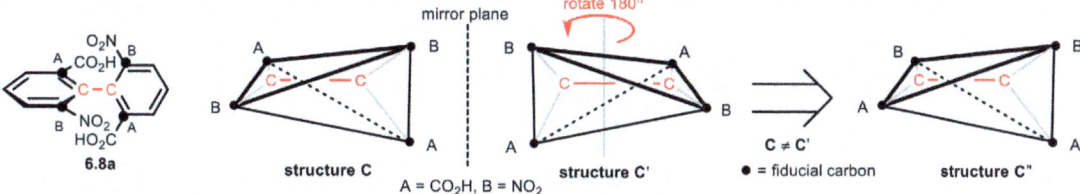

Figure 6.12   The rules for chirality in axially chiral molecules.

6.8a (red) makes up the stereogenic axis and the four *ortho* carbon atoms (shown with dots in compound 6.8a) carrying the functional groups CO_2H (represented as A) and NO_2 (represented as B) are positioned at the vertices of structure C. The carbon atoms represented by dots in compound 6.8a are sometimes called **fiducial groups.** The mirror image structure C′ is not superimposable on structure C.

❗ Key Learning Point

We are used to chiral molecules having stereogenic centres where all four groups have to be different, but in molecules with a chiral axis, such as biaryls, this rule is relaxed and as long as the single bond (*i.e.* C–C bond in red) lies on the axis of chirality, it's sufficient that *each* atom of the single bond has two different groups (*i.e.* A ≠ B).

❗ Key Learning Point

If you rotate structure C′ (Figure 6.12) 180° about the axis shown, you get structure C″. It is now clearer that structure C″ ≠ structure C, highlighting that they are enantiomers.

> Remember, for an axis to be stereogenic, only the groups attached to the atoms at the end of the axis need to be different.

6.9   Not All Biaryl Derivatives Are Chiral

We saw in Figure 6.11 that the atropisomers 6.8a and 6.8c of the hindered biaryl system were enantiomers of one another. These enantiomeric atropisomers can be separated by resolution. The atropisomers exhibit axial chirality (the stereogenic axis being through the C–C single bond) and there is no stereogenic atom. Before we carry on, we should provide the official IUPAC *Gold Book* definition for the stereogenic axis: '*An axis about which a set of ligands is held so that it results in a spatial arrangement which is not superimposable on its mirror image*'.[5] For biaryls (compound 6.9), this axis can be drawn through the C-4 → C-1 → C-1′ → C-4′ atoms (Figure 6.13). The fact that compounds 6.8a and 6.8c′ (Figure 6.11a) are non-superimposable mirror images of one another

6.9 | **6.10** achiral | **6.11** achiral

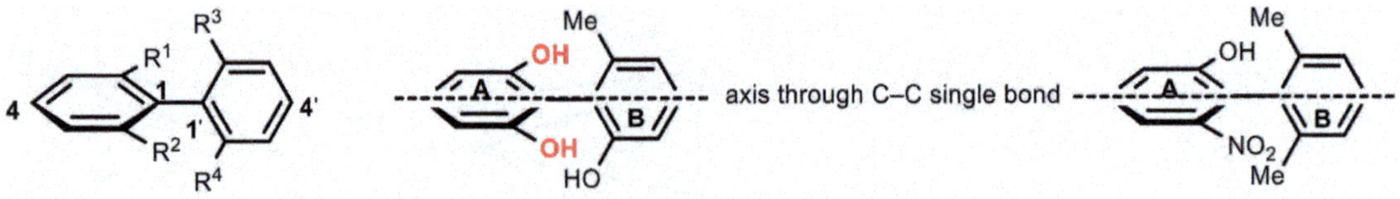

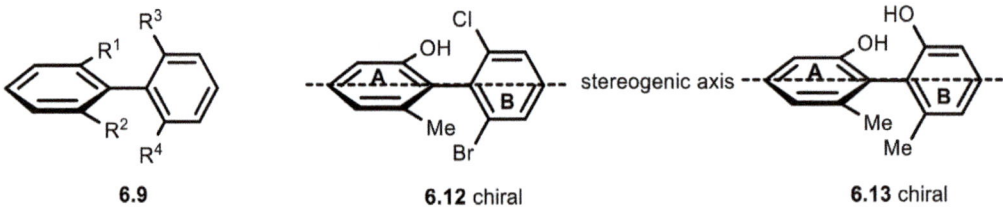

Figure 6.13 Achiral biaryl molecules.

> See the official IUPAC definition for this and lots of other terms: http://goldbook.iupac.org/terms/view/C01059

(enantiomers) can be difficult to visualise for some students and not all biaryls are chiral. The 'wordy' nature of the official IUPAC definition can also be intimidating. Luckily, there are some simple rules that will allow us to decide whether any biaryl derivative can exist as a pair of enantiomers or not. It involves inspecting the **fiducial** substituents on either side of the axis through the C–C single bond for *each* ring in turn.

6.9.1 Molecules With Ortho-substituents Only

Rule 1. If in a biaryl (compound 6.9) *either* the substituents $R^1 = R^2$ *or* $R^3 = R^4$, then the molecule is achiral.

An example of this is compound 6.10 (Figure 6.13), Ring A is symmetrical about the axis of the C–C single bond (both red *ortho* substituents are the same, namely hydroxyl groups). For compound 6.11, it is Ring B that is symmetrical around the axis drawn (Figure 6.13); compounds 6.10 and 6.11 are both achiral.

The converse of this is true, so we have Rule 2. If in the biaryl (compound 6.9), the substituents $R^1 \neq R^2$ *and* $R^3 \neq R^4$, the molecule is chiral.

Compounds 6.12 and 6.13 are both chiral (Figure 6.14). Compound 6.12 has different substituents on each side of the axis of the C–C single bond for Ring A and Ring B. Compound 6.13 has different substituents on each side of the axis of the C–C single bond for Ring A and Ring B *and* it doesn't matter that the substituents on Ring A (OH and Me) are the same as on Ring B (OH and Me)!

6.9 | **6.12** chiral | **6.13** chiral

Figure 6.14 Chiral biaryl molecules.

Let's check you've got this right. Decide whether molecules 6.14 to 6.16 in Figure 6.15 are chiral or not.

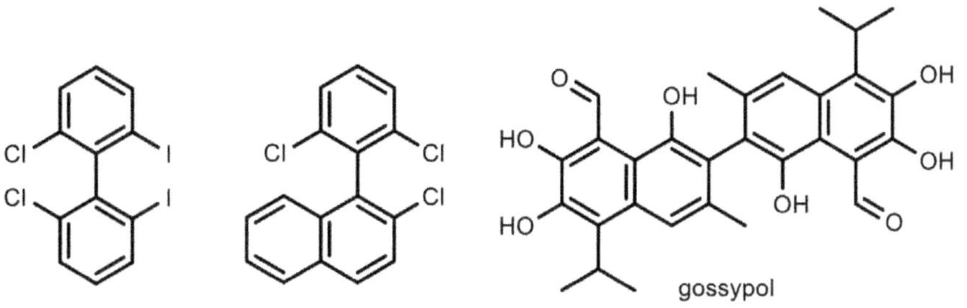

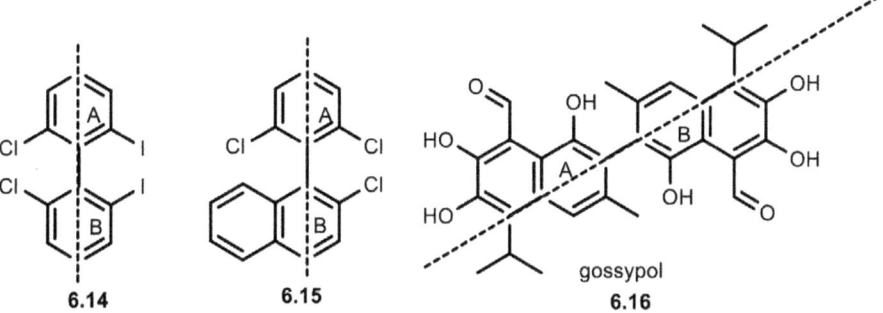

Figure 6.15   Which of these molecules is chiral?

Hopefully you identified that compound 6.14 is chiral. Both Ring A and Ring B (see Figure 6.16) have different groups either side of the C–C single bond axis, so Rule 2 applies. However, for compound 6.15, while Ring B has different groups either side of the C–C single bond axis (a chlorine on one side and a fused aromatic ring on the other), Ring A is symmetrical about the axis. Rule 1 applies and we can therefore be sure that 6.15 is achiral.

Figure 6.16   The biaryl axes for compounds 6.14 to 6.16.

Gossypol (compound 6.16) is a natural product that was evaluated as a male contraceptive. Don't get put off by the more complex structure. Since both Ring A and Ring B have different groups either side of the C–C single bond axis, Rule 2 applies. Gossypol is chiral. As expected, both enantiomers of gossypol have different biological activities.

6.9.2   Molecules With meta-substituents

It can get a bit more complicated if the derivatives have *meta* substituents as well. The key is to still make sure that each ring has a different arrangement of groups along either side of the axis through the C–C single bond. While all the *ortho* substituents are

> This is so important we are going to repeat it. As long as *both* substituents in Ring A are different *and* both substituents in Ring B are different, it doesn't matter if the two groups on Ring A (R^1 and R^2) are the same as those on Ring B.

the same on both Ring A (red) and Ring B (black) for the highly toxic polychlorinated biphenyl (compound 6.17) in Figure 6.17, the key is to look at the groups either side of the axis as a whole. Ring A also has a *meta* chlorine atom on one side of the axis but not on the other side at the *meta* position and the same can be determined for Ring B. Thus, neither Ring A nor Ring B is symmetrical about the axis. The two enantiomers (atropisomers) 6.17a and 6.17b are shown in Figure 6.17 and can be separated at room temperature.

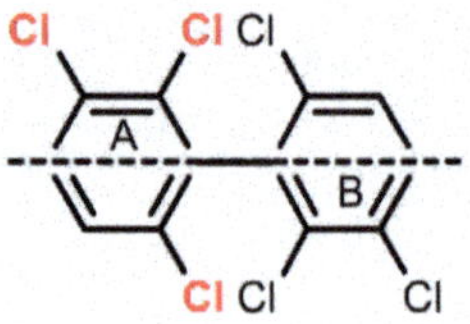

6.17 chiral

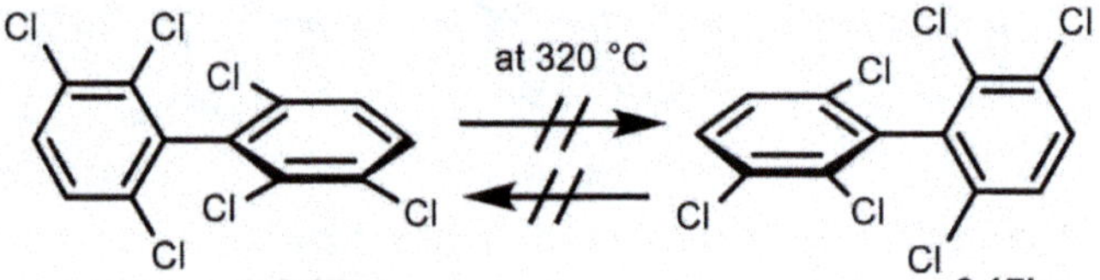

6.17a
6.17b
atropisomers stable at elevated temperature

at 320 °C

Figure 6.17   Chiral and achiral biaryl molecules.

It's probably good to check you've got this point too, by deciding whether compounds 6.18 and 6.19 in Figure 6.18 are chiral or not.

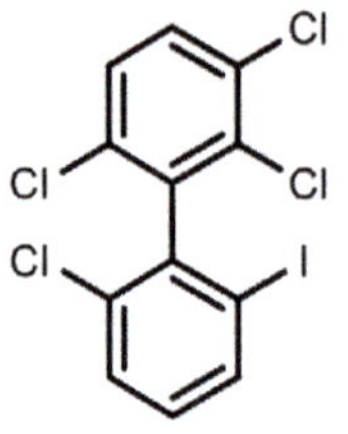

6.18

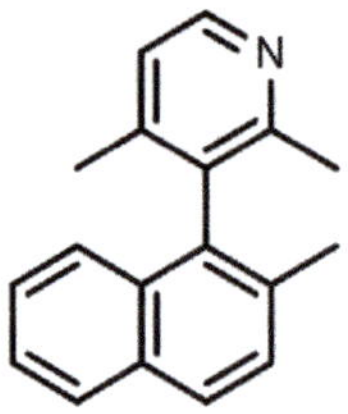

6.19

Figure 6.18   Are the molecules chiral?

For compound 6.18, although Ring A has two identical *ortho* chlorine substituents, the symmetry is broken at the *meta* substituent (Figure 6.19) as the H atom on the left (red) is different from the chlorine atom on the right (red). So, compound 6.18 is chiral. Compound 6.19

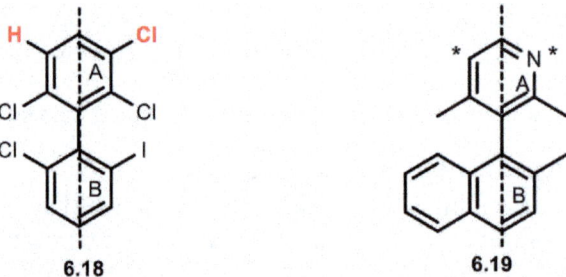

6.18

6.19

Figure 6.19   Compounds are asymmetrical around the stereogenic axis.

is chiral too – both Ring A and Ring B have different groups either side of the C–C single bond axis. Although Ring A has two identical *ortho* methyl substituents, the symmetry is broken with the pyridine ring nitrogen on one side of the axis (see the asterisks in Ring A of compound 6.19 in Figure 6.19).

6.10   Atropisomer Interconversion Barriers[6]

Evidence suggests that, with four *ortho* substituents, most biaryl derivatives, such as compound 6.17 (Figure 6.17), rotate so slowly that the corresponding atropisomeric enantiomers can be separated at room temperature. The barrier to interconversion of the enantiomers 6.17a and 6.17b (Figure 6.17) has been measured at 210 kJ mol^{-1} and no interconversion (racemisation) of the separated enantiomers was detected after 3 h, even at 320 °C.[6] A number of derivatives with three *ortho* substituents can also be separated (see compound 6.20a ↔ compound 6.20b in Figure 6.20) but, as expected, the barrier to rotation lowers with the decreased hindrance at the *ortho* positions. Consequently, very few biaryls with only two *ortho* substituents have had their atropisomeric enantiomers isolated at room temperature. Those that contain very bulky substituents, for example the isopropyl groups in compound 6.21a (Figure 6.20) have barriers to rotation around the central C–C bond close to the limit required to be able to successfully separate the atropisomers at room temperature (100 kJ mol^{-1}). Racemisation of a pure sample of atropisomer 6.21a was found to occur with a half-life of about 2 h at 80 °C.

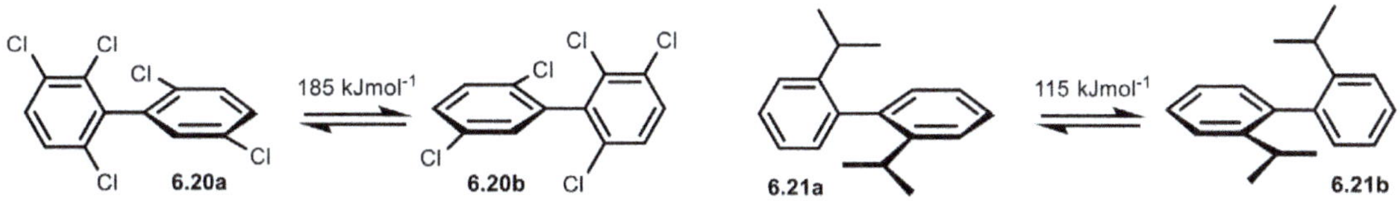

Figure 6.20   Interconversion of atropisomers is dependent upon steric hindrance at the *ortho* positions.

Again, let's check our understanding here. Firstly, think about whether compound 6.22 in Figure 6.21 is chiral or achiral. Would

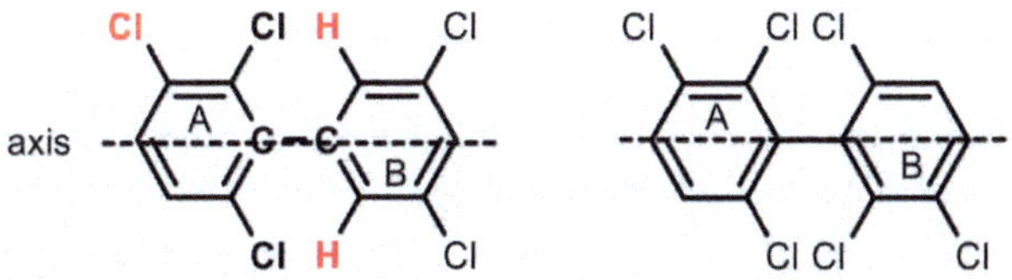

6.22 achiral or chiral?   **6.17**

Figure 6.21   Checkpoint question.

you expect the barrier to rotation around the central C–C bond in the biphenyl to be higher or lower than in compound 6.17?

Hopefully you spotted that compound 6.22 is achiral, because even though Ring A is not symmetric around the axis drawn, Ring B is symmetric, with both sides of the axis having one *ortho* H atom (red) and one *meta* chlorine atom. The rotation around the C–C bond in compound 6.22 is much faster than for compound 6.17, owing to the small *ortho* H substituents in Ring B, compared with the corresponding chlorine substituents in compound 6.17.

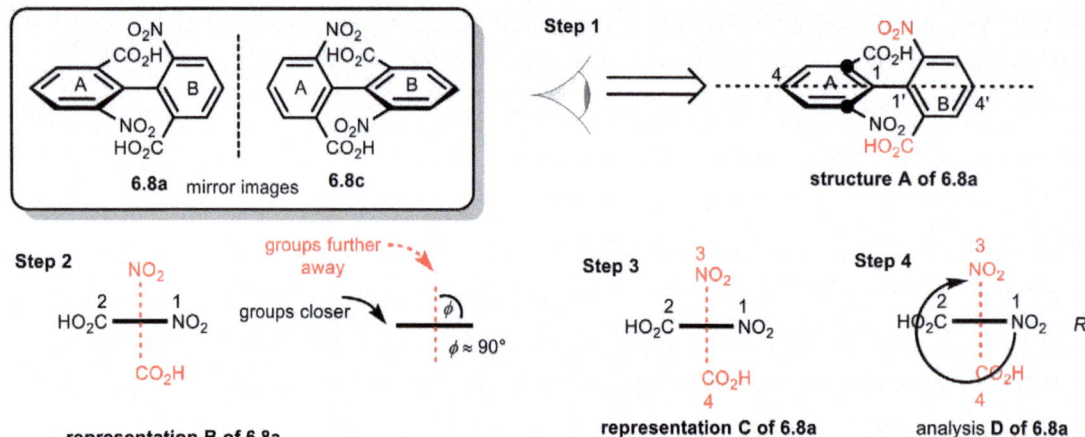

Figure 6.22   How to assign axial chirality as *R* or *S* in biaryl derivatives.

6.11   Axially Chiral Molecules: *R/S* Nomenclature

Assigning *R* or *S* configurations to axially chiral derivatives is a different challenge from assigning molecules with stereogenic atoms such as carbon (see Chapter 3). We will illustrate the process by considering the enantiomers 6.8a and 6.8c we met in Section 6.8.

Step 1. Firstly, we have to identify the stereogenic axis. For compound 6.8a, it is represented in Figure 6.22 (structure A) by the dotted line through the C-4–C-1–C-1'–C-4' axis. Then we have to look down the stereogenic axis (we can choose to look from either direction, as it doesn't matter). In this example, we are *choosing* to look from the left (see Figure 6.22, Step 1).

Step 2. We then draw a projection of what we see (remember that the two rings are twisted with respect to each other); for the analysis, we can regard the dihedral angle of the twist ϕ as 90°. We use the Cahn–Ingold–Prelog (CIP) rules in the normal way (for revision of these rules, see Chapter 2) to prioritise the fiducial groups (the *ortho carbon atoms*) of the ring nearer to us (ring A) (Figure 6.22, Step 2). In this case, the carbon with the NO_2 group (black) takes priority 1 and the carbon with the CO_2H group (black) takes priority 2. We draw a broad line for the groups near us (black line, *representation B*) and a dashed line to represent the groups attached to the ring further away from us (Ring B), in this case a red dotted line (Figure 6.22, Step 2). Note that if both *ortho* substituents are the same *and* hence both *ortho* carbons (the fiducial groups), we should continue around to the *meta* positions to determine priority (following the normal CIP rules).

Step 3. We then prioritise the groups on the ring further from us (Ring B, substituents in red) but take the higher group (the carbon with the NO_2) and label it 3rd priority and the lower group (the carbon with the CO_2H) and label it 4th priority (representation C, Figure 6.22).

Step 4. Ignoring the lowest-priority group (in this case the 4th-priority CO_2H group in red), we draw a curved line joining the 1st → 2nd → 3rd priority groups (black arrow). If this is clockwise, we assign the R configuration; if it is anticlockwise, we assign the S configuration. In this case, the configuration analysis assigns the molecule with the R configuration (analysis D, Figure 6.22).

Let's check that you can assign compound 6.8c (Figure 6.23) as the aS isomer.

We need to look down the stereogenic axis. It doesn't matter which way, but let's choose to look from the left, (Figure 6.24, Step 1). Ring

> Strictly speaking, the configurations should be written as aS or aR or sometimes Ra or Sa, where the lower case 'a' indicates that the configuration arises from a stereogenic *axis*, but a lot of textbooks often omit this, so be prepared to meet both.

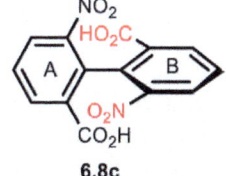

6.8c

Figure 6.23 Checking that compound 6.8c is the aS isomer.

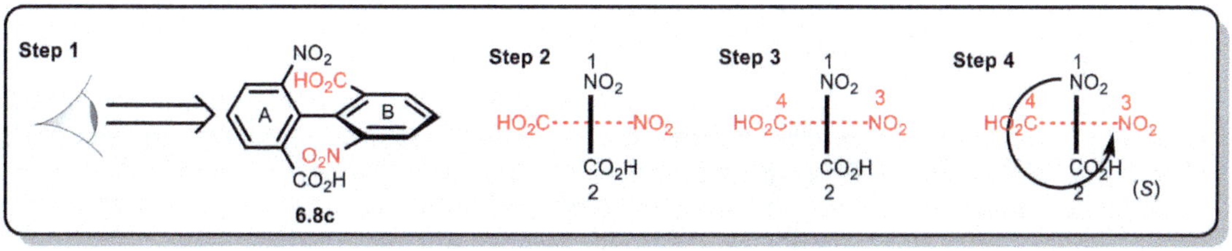

Figure 6.24 Assigning 6.8c as the aS isomer.

A is closest to us, so in Step 2 the groups NO_2 and CO_2H are added to a bold line. The two substituents in red (furthest away from us) are added as a dotted line with a dihedral angle of 90°. Prioritising using the CIP rules for the two groups closest to us gives highest priority (NO_2) and second-highest (CO_2H). When we do the same for the groups on the Ring B furthest from us (in red), the higher-priority group (NO_2) is given priority three and the lower-priority group (CO_2H) priority four (Step 3). Ignoring the lowest-priority group, we draw a curved line joining the 1st → 2nd → 3rd priority groups (black arrow) (Step 4). This is anticlockwise and we assign the *aS* descriptor.

6.12 Axially Chiral Molecules: *M/P* Nomenclature

Chemists may use the *aR* or *aS* configuration descriptors for axially chiral molecules but they can also use the *M* or *P* configuration assignment we met earlier for helicenes. The first two stages in assigning an *M* or *P* configuration are identical to that described for the *aR* and *aS* configuration descriptors. We will illustrate the process by considering enantiomer 6.8a (Figure 6.25).

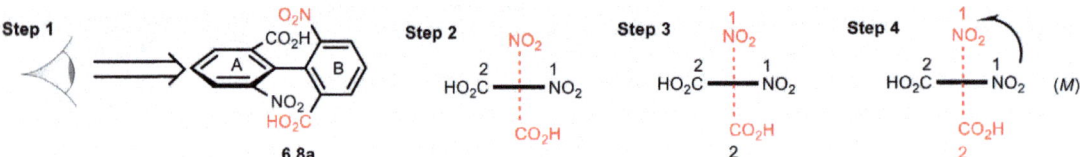

Figure 6.25   Assigning chirality using the *M* and *P* descriptors.

Step 1. Firstly we have to identify and look down the axis of chirality. In this example, we are *choosing* to look from the left of compound 6.8a (see Figure 6.25).

Step 2. We then draw a projection of what we see, and use the CIP rules in the normal way to prioritise the groups on the ring nearer to us (ring A). In this case, the NO_2 group takes priority 1 and the CO_2H group takes priority 2. We draw a broad line (black) for the groups near us and a dashed line to represent the groups further away from us (a red dotted line in this example).

Step 3. This is where the difference occurs between the approach described in Section 6.11 for the *aR* or *aS* configurations. For *M* or *P* assignment, we prioritise the groups on the ring further from us (Ring B) as before but label the higher-priority group 1 (the NO_2) and the lower-priority group 2 (the CO_2H in this case).

Step 4. We next draw a curved line joining the higher-priority group on the ring closer to us to the group with the higher priority on the ring further away from us. If this is clockwise, we assign the *P* configuration; if it is anticlockwise, we assign the *M* configuration.

Just as we did with the *aR* and *aS* configuration descriptors, let's reinforce our understanding of this different nomenclature by checking that compound 6.8c (Figure 6.26) has the *P* configuration.

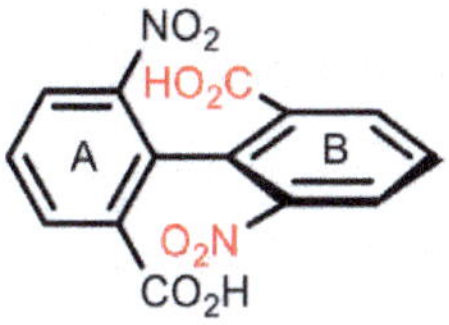

NO$_2$
HO$_2$C

A   B

O$_2$N
CO$_2$H

6.8c

Figure 6.26   Checking that compound 6.8c is the *P* isomer.

Remember, we first look down the stereogenic axis of compound 6.8c (Step 1, Figure 6.27). We choose to look from the left in this case. In Step 2, because Ring A is closest to us, the groups NO$_2$ and CO$_2$H are added to a bold line. The two substituents in red (further away from us) are added as a dashed line with a dihedral angle of 90°. Prioritising using the CIP rules for the two groups closer to us gives highest priority (NO$_2$) and second-highest (CO$_2$H). We do the same for the groups on the Ring B further from us (in red); the higher-priority group (NO$_2$) is given priority one and the lower-priority group (CO$_2$H) priority two (Step 3). Finally (Step 4), we draw a curved line joining the higher-priority group on the ring closer to us to the group with the higher-priority on the ring further from us (black arrow). This is clockwise and confirms that the answer is *P*.

> **◾ Checkpoint**
>
> You should now understand what makes biaryl derivatives axially chiral and be able to assign the configuration of enantiomers using either the *R/S* system of descriptors and the *M* and *P* descriptors.

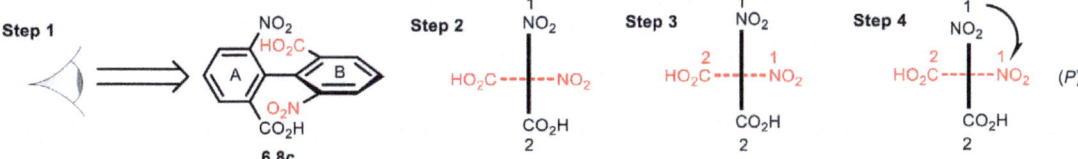

Figure 6.27   Assigning chirality of **6.8c** as *P*.

6.13 Bonding in Alkenes and Allenes[7]

You should remember that alkene functional groups are made up of a C–C σ bond and a C–C π bond where both carbon atoms are sp^2 hybridised (Figure 6.28); see Chapter 2 for a more detailed

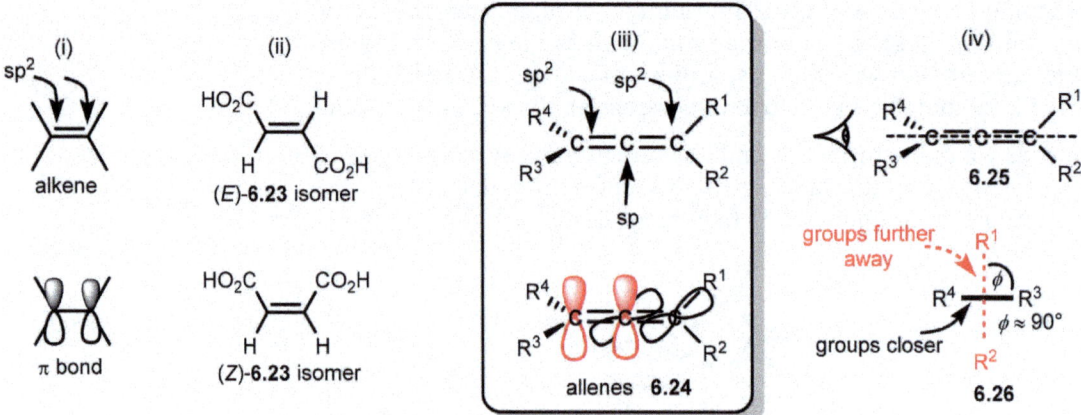

Figure 6.28   Structure of alkenes and allenes.

discussion. The consequence of this is that the alkene group is planar and rotation around the alkene is not possible without breaking the π bond. This leads to appropriately functionalised alkenes exhibiting isomerism (*e.g.* (*E*)-**6.23** and (*Z*)-**6.23**, (Figure 6.28ii)); see Chapter 2. The planar nature of the alkene functional group means that it is not in itself chiral. Conversely, the bonding in allenes (compound 6.24) (cumulated double bonds) is different (Figure 6.28iii)). The central carbon is sp hybridised and one of its p orbitals is used to form one alkene while the other **orthogonal** p orbital is used to furnish the other alkene. Thus, allenes are not planar and the two alkenes exhibit a dihedral angle of 90° with regard to one another, just like biaryls. This also places the substituents on each carbon atom orthogonal to each other (Figure 6.28iv), and if we look down the line of the axis of the three carbon atoms of allene 6.25 (arbitrarily from the left in this example) we could draw a projection to represent this (projection 6.26), as with biaryls. Thus, the broad bond in black represents the carbon atom and substituent nearer to us (R^3 and R^4 in this case) and the dashed bond (red) that which is further away (R^1 and R^2).

6.14   Axial Chirality in Allenes

Just as with biaryls, the orthogonal arrangement of four substituents coupled with the lack of rotation around the atoms in the chiral axis can lead to chirality. In other words, the allene substituents (R^1, R^2, R^3, R^4) are positioned at the vertices of a lengthened tetrahedron (Figure 6.29). The same rules governing axial chirality and the assignment of *aR/aS* or *M/P* stereochemical descriptors in biaryls will govern allenes (see Sections 6.11 and 6.12). For allenes,

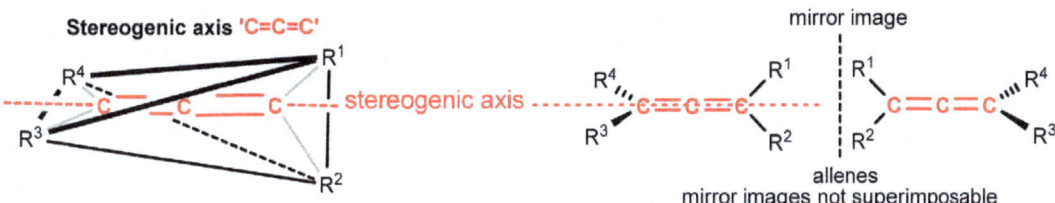

Figure 6.29   Elongation of a tetrahedron leads to a distorted tetrahedron with a stereogenic axis.

the stereogenic axis is the line running through each of the three carbon atoms of the allene, and as long as $R^1 \neq R^2$ *and* $R^3 \neq R^4$, the molecule will be chiral.

❗ Key Learning Point

The orthogonal π orbitals in allenes can lead to axial chirality.

Let's test our understanding of the chirality in allenes by considering which of the allene molecules 6.27 to 6.30 in Figure 6.30 are chiral.

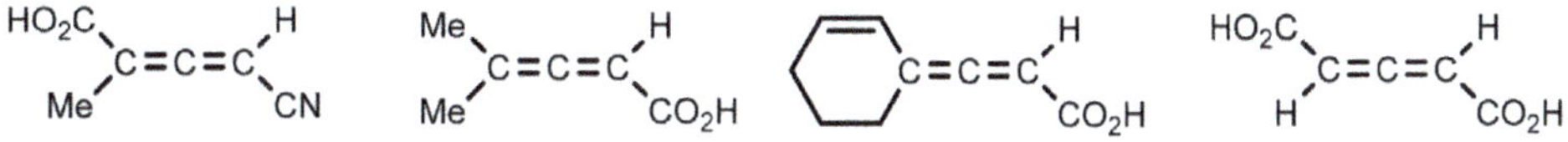

HO$_2$C$\diagdown$ $\diagup$H C=C=C Me$\diagup$ $\diagdown$CN	Me$\diagdown$ $\diagup$H C=C=C Me$\diagup$ $\diagdown$CO$_2$H	⬡C=C=C$\diagup$H $\diagdown$CO$_2$H	HO$_2$C$\diagdown$ $\diagup$H C=C=C H$\diagup$ $\diagdown$CO$_2$H
6.27	**6.28**	**6.29**	**6.30**

Figure 6.30   Which of these allenes are chiral?

Remember that, as with biaryls, we must first draw the stereogenic axis through the allene functional group (dotted line), as shown in Figure 6.31. Then for the carbon at each end of the allene in turn (C-1 and C-3), we must determine whether the substituents are the same or different. For the molecule to be chiral, both groups on C-1 and C-3 must be different *but* remember *it doesn't matter if the two different groups on C-1 are the same as the two different groups on C-3 (this will become obvious if not so now!).*

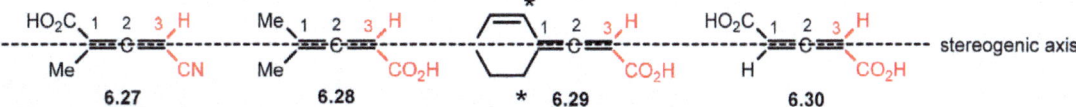

Figure 6.31   Assessing whether compounds 6.27 to 6.30 are chiral.

Compound 6.27. C-1 has two different groups (CO$_2$H and Me), and C-3 has two different groups (H, CN), so the molecule is chiral.

Compound 6.28. C-1 has two groups identical (Me and Me) so, despite C-3 having two different groups (H and CO_2H), the molecule is achiral.

Compound 6.29. C-1 has two different groups (note that the two starred atoms are different and the groups either side of the stereogenic axis for C-1 are not the same), and C-3 has two different groups (H and CO_2H), so the molecule is chiral.

Compound 6.30. C-1 has two different groups (H and CO_2H) and C-3 has two different groups (H and CO_2H), so the molecule is chiral. It doesn't matter that the two different groups on C-1 are the *same* as the two different groups on C-3.

The same rules that apply when assigning *aR/aS* and *M/P* stereochemical descriptors to biaryls also apply to allenes and all molecules with axes of chirality. We will illustrate this by assigning the *aR/aS* and *M/P* stereochemical descriptors to the allene 6.30 (Figure 6.32).

> 💬 Note that, as in Figure 6.9, the terminal carbon atoms of the allene (those attached to R^1, R^2, R^3 and R^4) are sp^2 hybridised.

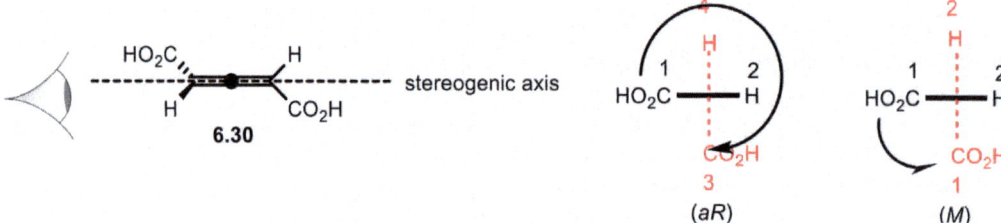

Figure 6.32 Working out the configuration of the allene 6.30.

Looking down the stereogenic axis from the left (arbitrary choice) and applying the rules from Sections 6.11 and 6.12, we obtain the priorities CO_2H = 1, H = 2 for the groups on the closer carbon to the eye (black), and CO_2H = 3, H = 4 for the groups further away (dotted red line). This confirms the *aR* configuration (clockwise, 1st → 2nd → 3rd). Using the *M/P* nomenclature, we assign the groups on the carbon further away CO_2H = 1, H = 2 (red). The arrow linking the higher-priority group on the carbon closer to us to that on the carbon further away is anticlockwise, so is assigned the *M* configuration.

> �◾ **Checkpoint**
>
> You should now understand what makes allene derivatives axially chiral and learn to assign the configuration of enantiomers using either the *R/S* system of descriptors and the *M* and *P* descriptors.

6.15 Importance of Axial Chirality in Organic Synthesis, Nature and Medicine[8–10]

At the start of the chapter, we introduced the structure of BINAP (compound 6.2), a bidentate phosphine ligand first introduced in the 1980s by Noyori. Hopefully by now you can appreciate that it has an axis of chirality (it is a biaryl derivative), owing to the restricted rotation in the binaphthyl portion facilitated by the large *ortho* phosphine substituents. The angle between the two naphthyl groups (dihedral angle) in compound 6.2 has been measured at 90°, indicating that the fiducial groups on each ring are orthogonal. It can be prepared and resolved by a range of synthetic routes and it is useful in catalysts for asymmetric synthesis. For example, when coordinated to ruthenium halides, (*S*)-BINAP (compound 6.2a) can reduce ketones in the presence of H₂ to give chiral alcohols in high enantioselective excess (Figure 6.33). A range of BINAP derivatives, such as SEGPHOS (compound 6.31), have been prepared and used in a range of different asymmetric synthesis applications, often with more selective results than BINAP. Consequently, chiral catalysts derived from biaryl derivatives have proven to be excellent tools in the chemists' toolkit to prepare other chiral molecules in asymmetric synthesis.

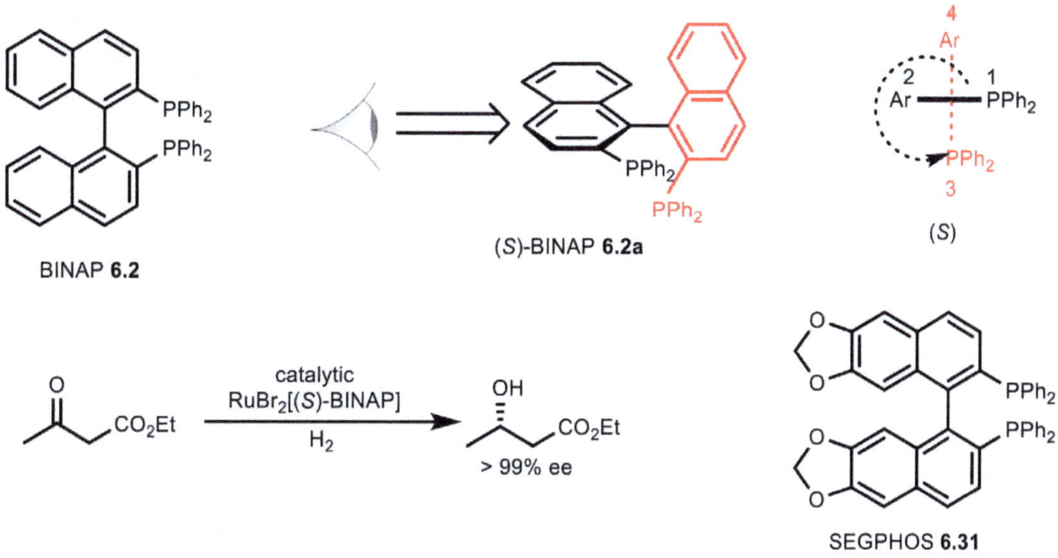

Figure 6.33 The structure and reaction of BINAP.

You have already been introduced to a number of naturally occurring molecules that exhibit axial chirality, namely the allene antibiotic mycomycin (compound 6.1 in Figure 6.1) and the potential male contraceptive biaryl gossypol (compound 6.16 in Figure 6.15). Restricted rotation in molecules leading to axial chirality is not restricted just to the C–C bonds found in allenes or in biaryls such as gossypol (compound 6.16) and BINAP (compound 6.2). Another single bond that can show slow rotation if suitably hindered (and hence potentially lead to atropisomers with axial chirality) is the C–N bond. The potential glycine transport inhibitor (GlyT1) (*R*)-6.3 exhibits slow rotation around the aryl C–N bond with a half-life of racemisation of 21 years at normal human body temperature (Figure 6.34). The slow rotation around the C–N bond ((*R*)-6.3 → (*S*)-6.3) is caused by the steric clash between the methyl group of the aromatic ring and the ethyl group of the triazole, which precludes the molecule becoming planar. Instead, there is a twist in the molecule around the C–N bond (just as in biaryls), leading to axial chirality.

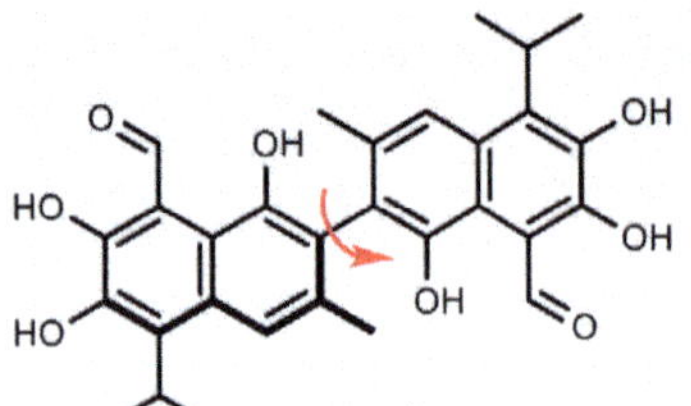

Figure 6.34 Class 3 atropisomers of potential drugs must be evaluated independently for biological activity.

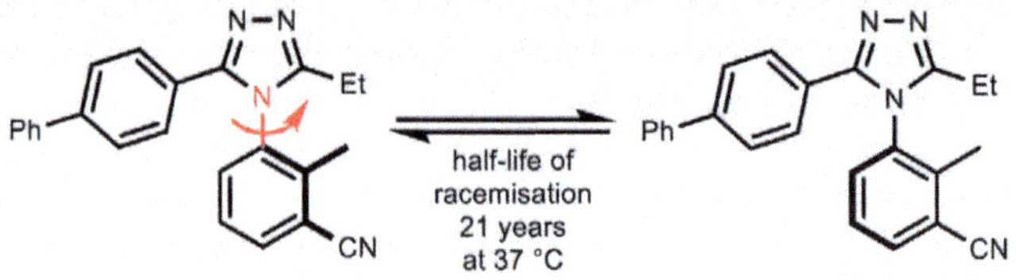

The carbon of the C–N bond is normally a sp² hybridized carbon atom.

Just as different enantiomers of molecules containing stereogenic centres will often exhibit different biological effects for each enantiomer (such as the drugs thalidomide and ibuprofen), the same is true for potential drugs that have an axis of chirality. In medicinal chemistry, atropisomers are divided into three categories. Class 1 atropisomers are molecules where barriers to rotation are low and isomers rapidly interconvert (in seconds) at room temperature. Potential drugs in this class can be evaluated as racemates. Class 2 atropisomers have intermediate stability, of the order of minutes to weeks, while class 3 atropisomers (like compound 6.3) are stable over a period of years with barriers to rotation above 125 kJ mol⁻¹. Owing to this stereochemical stability, molecules exhibiting class 3 behaviour should undergo preclinical testing as single stereoisomers.

6.16 Conclusion

Well done! This is a really tricky area of stereochemistry but, as you've seen, very important in modern organic chemistry. Hopefully, you now understand how molecules *without* a stereogenic atom can be chiral and how axial chirality arises, and you can now define what an atropisomer is.

In terms of your technical ability, you should now feel comfortable with assigning the configuration of helically chiral molecules using the *M* and *P* descriptors and of biaryl derivatives and allene derivatives using either the *R/S* system of descriptors or the *M* and *P* descriptors.

Finally, watch out! Helical chirality will pop up again in our final chapter, on inorganic compounds.

Exercises

Questions 1 to 3 test general understanding of Sections 6.10 to 6.14 in a format you might come across in a typical exam question. Questions 4 and 5 cover all the areas covered in this chapter and are more applied. Question 5 covers material covered in Chapter 3 as well as this chapter.

1. Which of the following molecules are chiral?

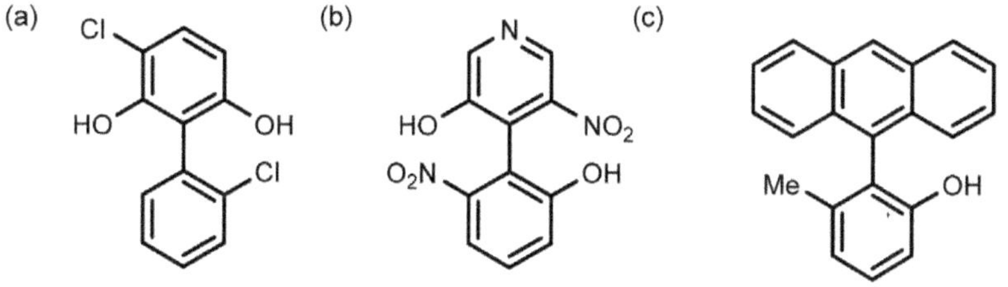

2. Draw one enantiomer of each of molecules 6.32 and 6.33.
   i. Assign the configuration of the enantiomer you have drawn as either *aS* or *aR*.
   ii. Assign the configuration of the enantiomer you have drawn as either *M* or *P*.

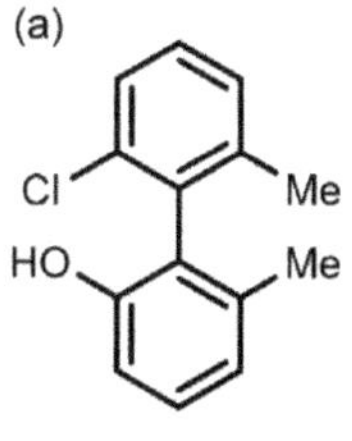

6.32 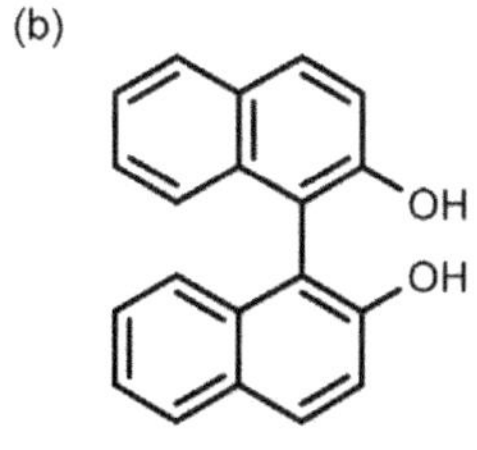   6.33

3. Using the *M/P* nomenclature, assign the absolute configuration of:
   (a) Allene 6.34
   (b) The natural product (+)-knipholone.

(a)

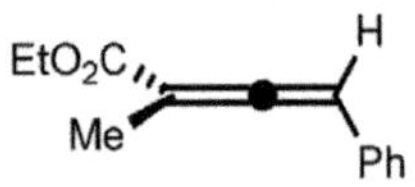

EtO$_2$C,,,

Me

H

Ph

6.34

(b)

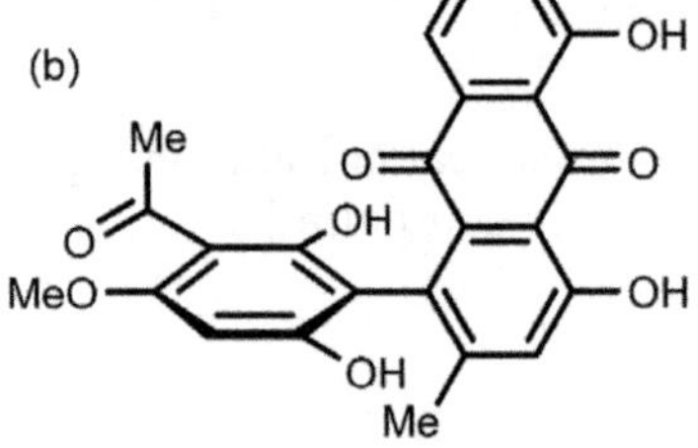

Me

MeO

O

OH

OH

Me

OH

O

O

OH

OH

(+)-knipholone
antimalarial

4. It is often possible that axial chirality can be transformed into helical chirality during the course of a reaction. The axially chiral atropisomer 6.35, when reacted with TiCl$_4$ furnished the thiaheterohelicene 6.36. Assign the axial chirality in compound 6.35 and the helical chirality in compound 6.36. (Hint: The two carbon atoms highlighted by dots indicate the C–C bond where restricted rotation occurs in compound 6.35.)

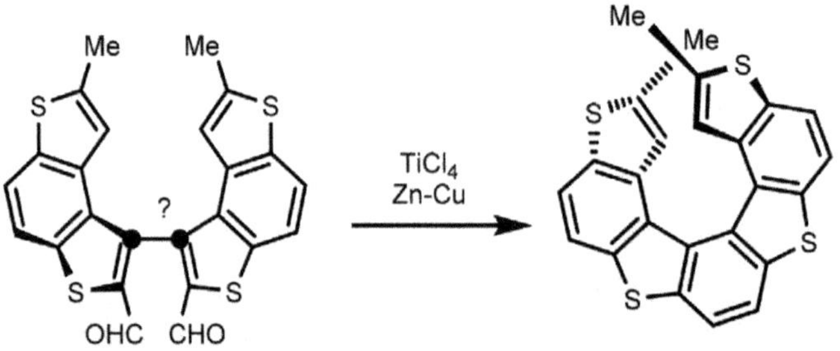

Me   Me

S   S

?

S   S

OHC   CHO

6.35

TiCl$_4$
Zn-Cu

Me   Me

S   S

S

S

S

6.36

5. Compound 6.37 contains a chiral centre (dotted carbon) and a chiral axis (C–N single bond). The large *ortho* iodine substituent restricts rotation around the aryl–N bond (arrow in red) and the aromatic ring is orthogonal to the amide. For compound 6.37, assign the two stereogenic elements using the *R/S* notation for the stereogenic centre and the *M/P* notation for the stereogenic axis. Draw the enantiomer and the two possible diastereoisomers of compound 6.37.

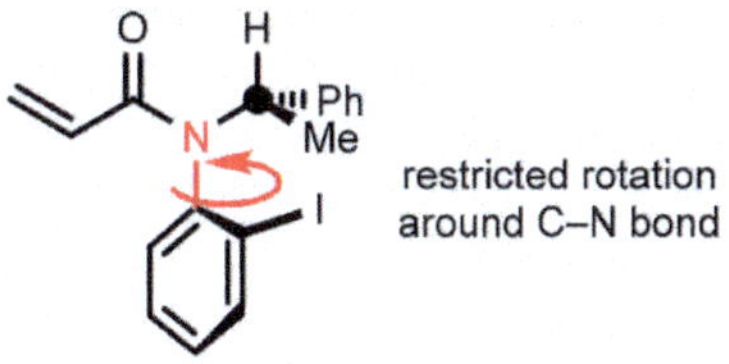
O
H
Ph
Me
N
I
restricted rotation
around C–N bond

6.37

Example Exam Question

This example question is chosen to test your understanding of the concepts covered in this chapter.

(a) Explain why the antibiotic mycomycin (compound 6.1) can exist as a pair of enantiomers, highlighting the functional group that is responsible for this.
(b) Draw both enantiomers of mycomycin (compound 6.1) and, using the Cahn–Ingold–Prelog rules, assign the absolute configuration of each enantiomer as *R* or *S*.
(c) Assign the configuration of the naturally occurring *R* enantiomer using the *M* and *P* nomenclature.

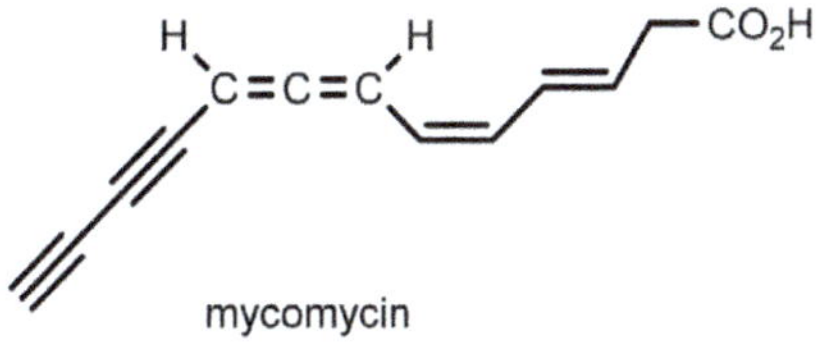
H
C=C=C
H
CO_2H

mycomycin
6.1

Answers to Exercises

1.

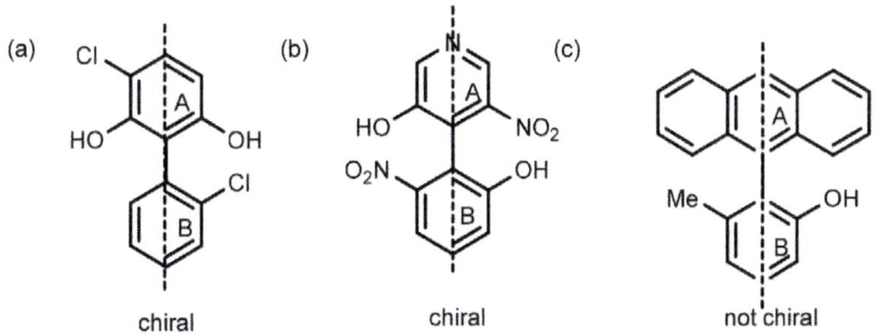
(a)
Cl
HO
OH
A
Cl
B
chiral
(b)
N
HO
A
NO_2
O_2N
OH
B
chiral
(c)
A
Me
OH
B
not chiral

2.

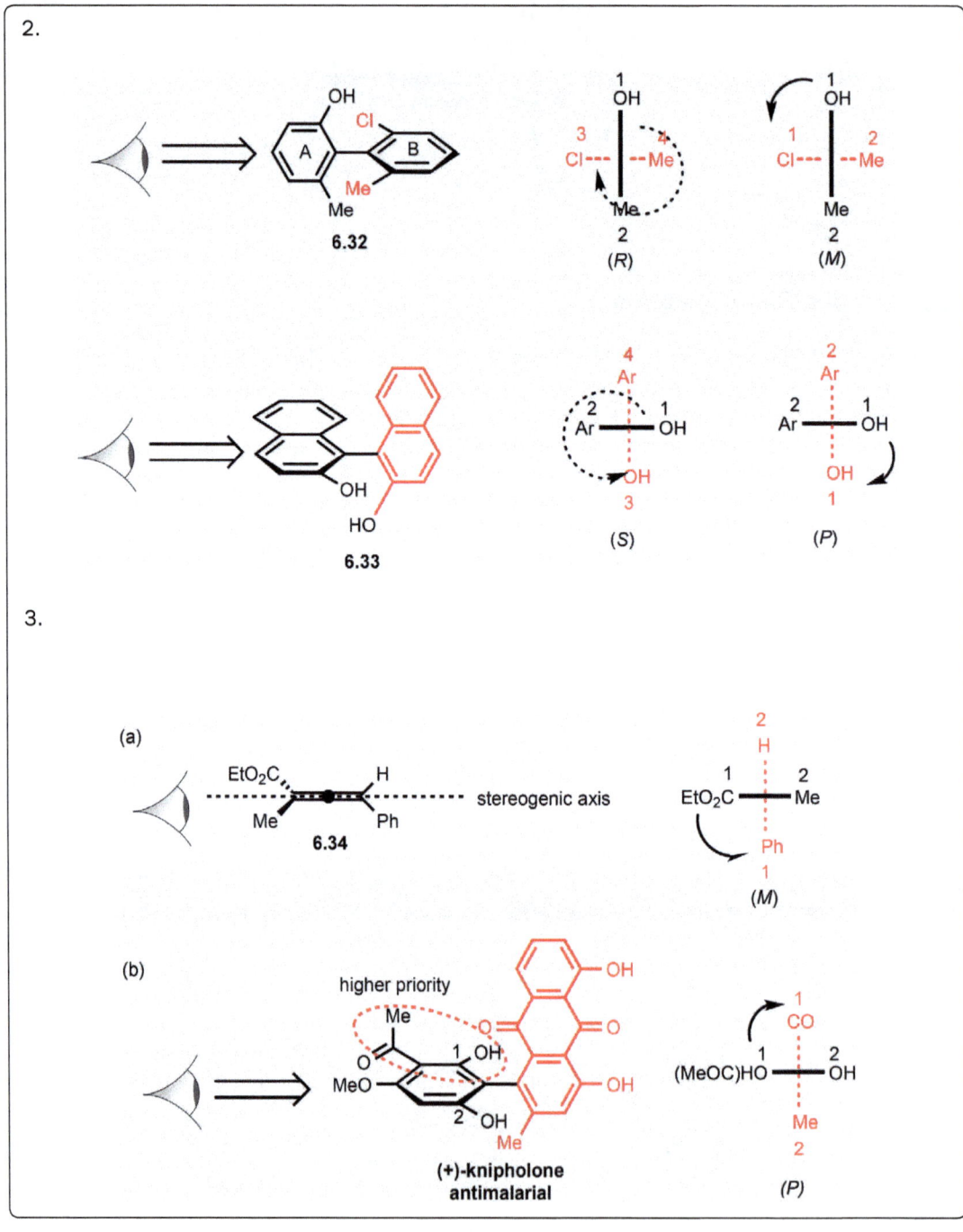

3.

(a)

stereogenic axis

6.34

(M)

(b)

higher priority

(+)-knipholone
antimalarial

(P)

4.

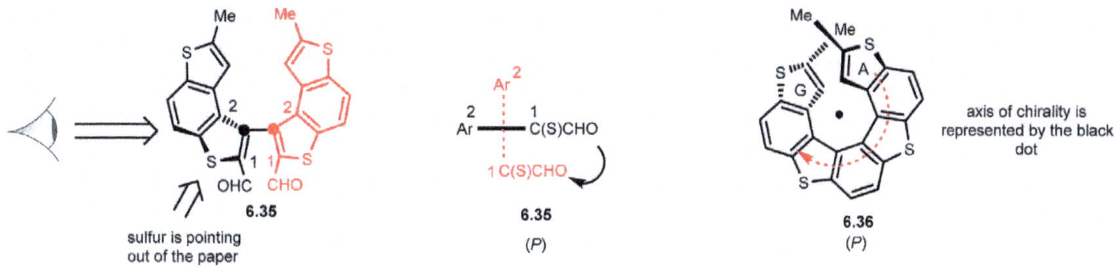

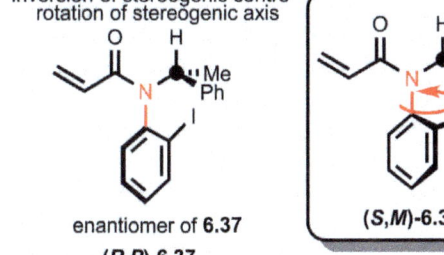

Me
Me
S
S
2
2
S
1
1
S
OHC
CHO
6.35
sulfur is pointing out of the paper
Ar 2
2 Ar
1 C(S)CHO
1 C(S)CHO
6.35
(P)
Me
Me
S
S
A
G
S
S
6.36
(P)
axis of chirality is represented by the black dot

5.

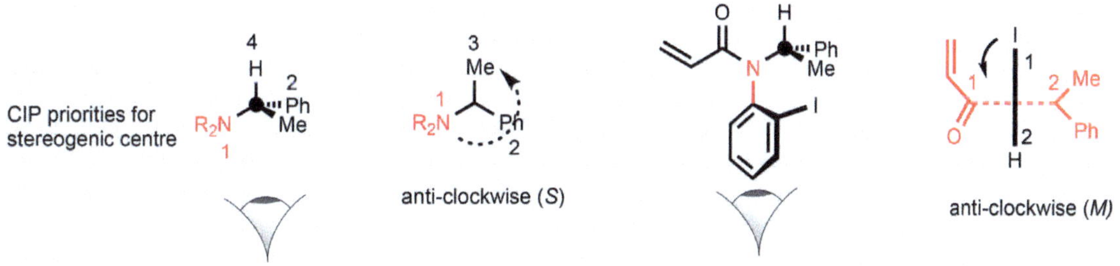

CIP priorities for stereogenic centre
4
H
2
Ph
Me
R_2N
1
3
Me
1
R_2N
Ph
2
anti-clockwise (*S*)
O
H
Ph
Me
N
I
I
1
1
2 Me
O
2
Ph
H
anti-clockwise (*M*)

The enantiomer of (*S*,*M*)-6.37 will have all elements of chirality reversed (*i.e.* (*R*,*P*)-6.37), while the diastereoisomers of (*S*,*M*)-6.37 will only have one element of chirality reversed (*i.e.* either (*S*,*P*)-6.37 or (*R*,*M*)-6.37).

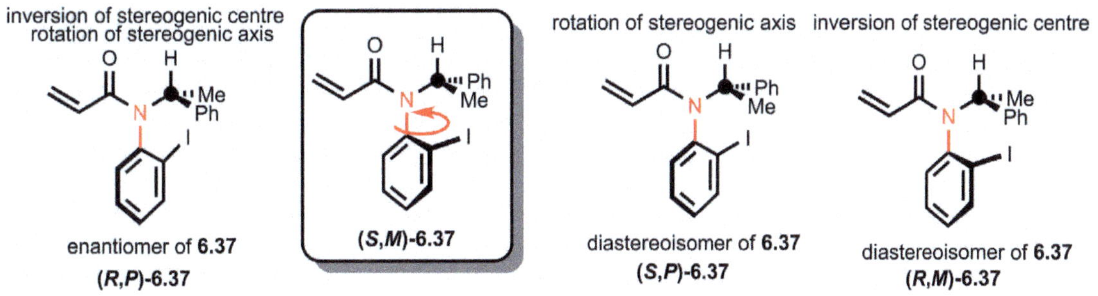

inversion of stereogenic centre
rotation of stereogenic axis
enantiomer of **6.37**
(*R*,*P*)-6.37
(*S*,*M*)-6.37
rotation of stereogenic axis
diastereoisomer of **6.37**
(*S*,*P*)-6.37
inversion of stereogenic centre
diastereoisomer of **6.37**
(*R*,*M*)-6.37

❓ Answer to Example Exam Question

(a) Owing to the bonding in the allene functional group, both pairs of substituents on the terminal carbons of the allene are orthogonal to one another; a stereogenic axis will lead to chirality as a result of this functional group. The mirror images are not superimposable.

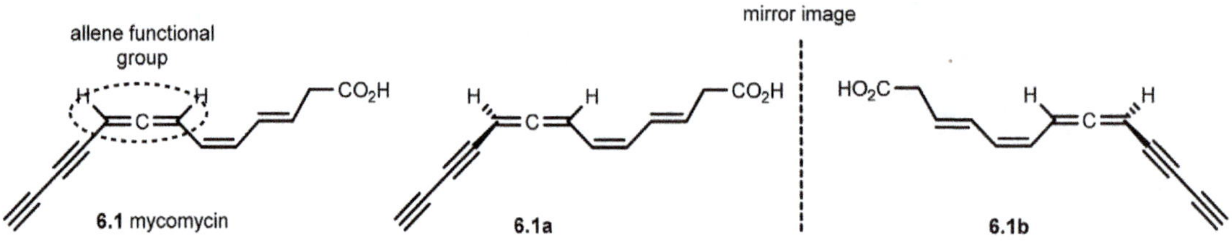

(b) For compound 6.1a, if we look down the stereogenic axis, in this case from the left, and apply the rules from Section 6.14. We assign the priorities alkyne substituent = 1, H = 2 for the groups on the closer carbon to the eye (black), and alkene substituent = 3, H = 4 for the groups further away (dotted red line). This leads to the aS configuration (anticlockwise, 1st → 2nd → 3rd). For compound 6.1b, the priority is clockwise, the aR configuration.

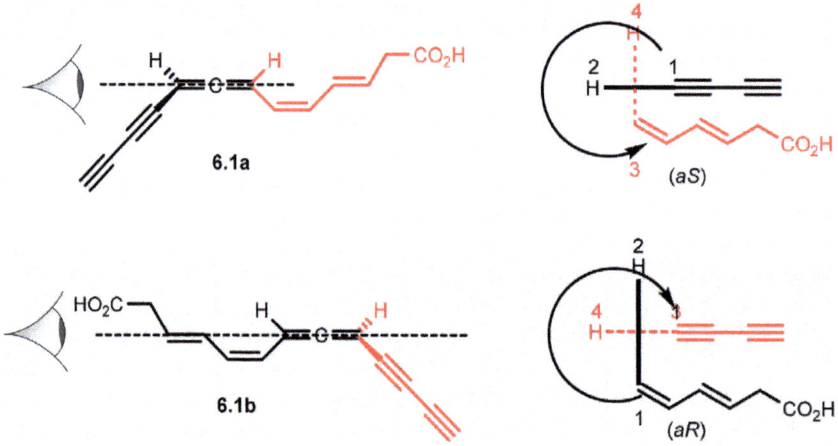

(c) Structure 6.1b was assigned as the *aR* enantiomer in part (b). Thus, using the *M/P* nomenclature, we assign the groups on the carbon further away, alkyne group = 1, H = 2 (red). The arrow linking the higher-priority group on the carbon closest to us to that on the carbon further away is anticlockwise, so assigned as the *M* configuration.

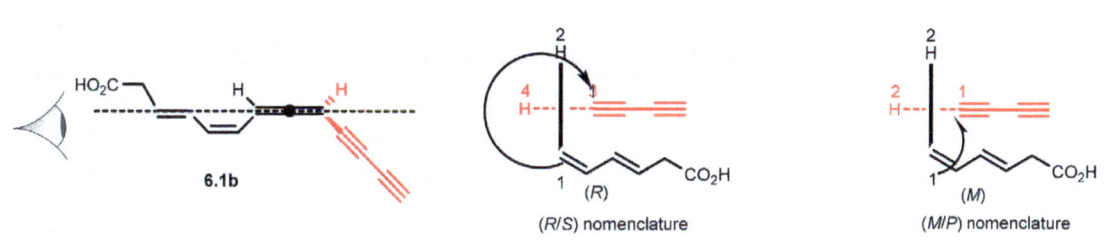

References

1. M. Gingras, *Chem. Soc. Rev.*, 2013, **42**, 968–1006.
2. A. Almenningen, O. Bastiansen, I. Fernholt, B. N. Cyvin, S. J. Cyvin and S. Samdal, *J. Mol. Struct.*, 1985, **128**, 115–125.
3. B. Testa, *Helv. Chim. Acta*, 2013, **96**, 351–374.
4. G. Bringmann, T. Gulder, T. A. M. Gulder and M. Breuning, *Chem. Rev.*, 2010, **111**, 563–639.
5. See http://goldbook.iupac.org/html/C/C01059.html, accessed 13/2/2019.
6. M. T. Harju and P. Haglund, *J. Anal. Chem.*, 1999, **364**, 219–223.
7. J. Ye and S. Ma, *Org. Chem. Front.*, 2014, **1**, 1210–1224.
8. R. Noyori and H. Takaya, *Acc. Chem. Res.*, 1990, **23**, 345–350.
9. P. W. Glunz, *Bioorg. Med. Chem. Lett.*, 2018, **8**, 53–60.
10. D. Campolo, S. Gastaldi, C. Roussel, M. P. Bertrand and M. Nechab, *Chem. Soc. Rev.*, 2013, **42**, 8434–8466.

By the End of This Chapter You Will:

☐ Understand stereochemistry in square planar complexes and be able to name *cis* and *trans* isomers.
☐ Be able to recognise and name *cis* and *trans* and *mer* and *fac* isomers of octahedral complexes.
☐ Be able to assign Δ or Λ configurations to chiral octahedral complexes with bidentate ligands.

What You Will Get from This Chapter

In Chapter 7, we'll draw on some key concepts from earlier chapters, including *cis* and *trans* relationships from Chapter 2, axial and equatorial positions from Chapter 5 and helical chirality from Chapter 6. So it's important to revisit those concepts before studying the stereochemistry of inorganic complexes. And if your molecular model kit doesn't have an octahedral atom with six points, it's time to borrow one!

Stereochemistry of Inorganic Molecules

7.1 Inorganic Molecules

So far, we've looked at stereochemistry from the point of view of compounds with carbon atoms. We've seen that bonds from carbon to other atoms can form linear, planar trigonal or 3D tetrahedral shapes that have consequences for the stereochemistry of organic chemicals.

As we move down the periodic table, we encounter elements that, because of their larger atomic radius or the different orbitals available to bond to other atoms, can have larger coordination numbers and different shapes. This includes compounds of main-group non-metals, such as phosphorus and sulfur (see Chapter 3), but our focus here will be on transition metal complexes.

Four-coordinate metal complexes can be tetrahedral. For these, everything we have learnt about the stereochemistry of sp^3 hybridised carbon centres applies. However, when d orbitals are involved, it is possible for the four ligands to make a **square planar** shape instead. A great example of this kind of complex is one of the most famous and important chemotherapy drugs, cisplatin, which is on the World Health Organization list of essential medicines.

7.2 Stereochemistry of Square Planar Complexes

Cisplatin has an isomer *trans*-platin (Figure 7.1), which is much less potent as a drug. Though the 'platin' part of these names is not an official IUPAC description, the *cis* and *trans* prefixes are the correct way to describe the stereochemical arrangement of ligands in such square planar complexes. So, we don't need to worry about

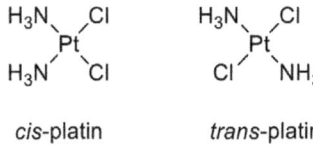

cis-platin　　　　*trans*-platin

Figure 7.1　Isomers cisplatin and *trans*-platin.

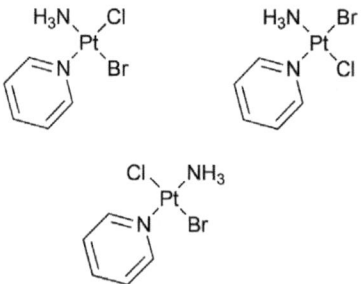

cis-platin	*cis*-but-2-ene (Z-but-2-ene)	*trans*-platin	*trans*-but-2-ene (E-but-2-ene)

Figure 7.2   Comparison of *cis*- and *trans*-platins and *cis*- and *trans*-but-2-enes.

Cahn–Ingold–Prelog (CIP) rules here. We just describe the relative stereochemistry we see. In cisplatin the ammine (NH_3) ligands are *cis* to each other, as are the chlorides, just like alkenes (Figure 7.2).

We still use *cis* and *trans* with MA_2BC complexes (M is a metal, the others are ligands), but we just describe the relative stereochemistry of the A ligands. Let's check you've got this, by drawing out and labelling the two isomers of square planar $Pt(NH_3)_2BrCl$.

Hopefully, you just focused on the two NH_3 ligands. They can either be *cis* or *trans* to each other (Figure 7.3).

cis isomer	*trans* isomer

Figure 7.3   *Cis*- and *trans*-MA_2BC complexes.

For completeness, let's mention MABCD complexes. All the ligands are different and there are three possible isomers (*e.g.* Figure 7.4).

Figure 7.4   Example of isomeric MABCD complexes.

> **◼ Checkpoint**
>
> You should now understand stereochemistry in square planar compounds and be able to name *cis* and *trans* isomers.

If you are interested, the conventions for naming these fairly rare complexes can be found in the IUPAC Red Book. Just to check you are visualising the compounds correctly, can you pick out the isomer with *trans* halide ligands?

Well done if you spotted the *trans* halides correctly (Figure 7.5).

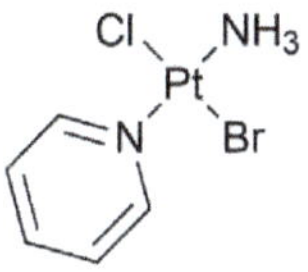

Figure 7.5   An isomer of an MABCD complex with *trans* halide ligands.

> The stereochemical analogy between square planar complexes and alkenes is useful but don't push it too far! Alkenes have two distinct 'ends' whereas platins do not. So *E/Z* nomenclature only has meaning in the alkene context.

7.3 Shapes of Five- and Six-coordinate Complexes

Now let's venture beyond four-coordinate. Five-coordinate complexes are common. They can have either '**trigonal bipyramidal**' or '**square pyramidal**' shapes (Figure 7.6). This mainly depends on what the metal M and the ligands L are, although some complexes switch from one to the other when their surroundings change!

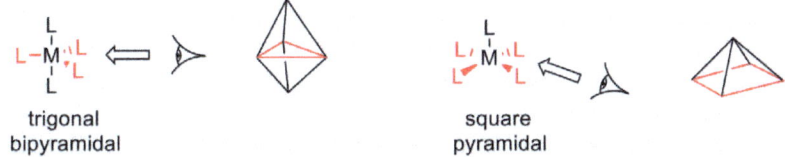

trigonal
bipyramidal

square
pyramidal

Figure 7.6   Five-coordinate ML_5 complexes can be trigonal bipyramidal or square pyramidal.

The only stereochemical feature of five-coordinate complexes we're going to worry about in this book is how to identify ligands as being in either 'axial' (a) or 'equatorial' (e) positions. Hopefully, the comparison with the positions of C–H bonds in cyclohexane (see Chapter 5) is helpful in spotting the axial and equatorial sites (Figure 7.7).

> If you can't easily see the pyramidal shapes in the chemical structures, try joining up different sets of three ligands L into triangles and you should be able to identify the faces of the pyramids (except for the base of the square pyramid, which is, of course, a square!).

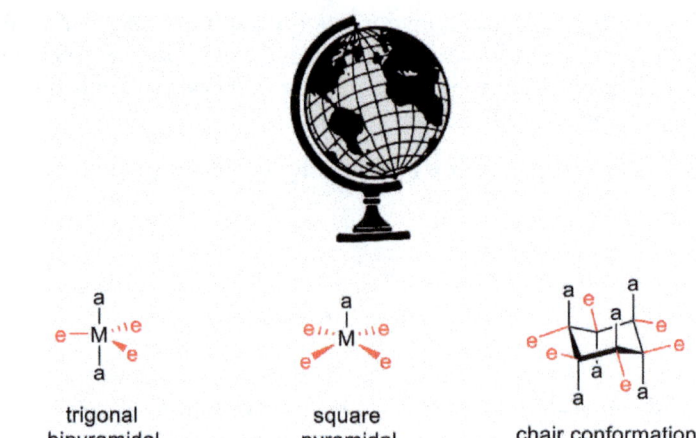

Figure 7.7   Axial (a) and equatorial (e) positions in five-coordinate complexes and cyclohexane.

Let's now look at a six-coordinate complex where all six ligands are the same. A classic example is hexaamminecobalt(III) chloride, with the formula $[Co(NH_3)_6]Cl_3$. Each ammine (NH_3) ligand occupies an equivalent site at an apex of a regular octahedron. There are no axial or equatorial positions here. Because of the high symmetry, all six octahedral positions are identical (Figure 7.8). With six identical ligands, all bond lengths and angles are the same. If your molecular model kit has a six-coordinate atom in it, why not make a representation of this cation? It'll come in useful in the next section.

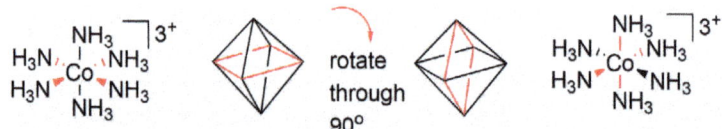

Figure 7.8   Rotating an octahedral ML_6 complex shows us that all six vertices are equivalent.

7.4   Stereochemistry of Achiral Six-coordinate Complexes

If you've made a model of the octahedral trication in Figure 7.8, now try removing two of the ammine ligands and replacing them with chlorides to make $[CoCl_2(NH_3)_4]^+$. What possibilities are there for placement of the Cl ligands? How might you name your new complexes in terms of stereochemistry? Remember you're starting with a completely symmetrical octahedron, so it really doesn't matter which ligand you exchange first. They are all equivalent.

Hopefully, you'll have found that for an MA_2B_4 (generic formula) complex like this, the situation is very much like cisplatin and *trans*-platin. There are two possibilities and we do use *cis* and *trans* to describe the stereochemistry (Figure 7.9).

> This task gets much more obvious once you've replaced one ligand.

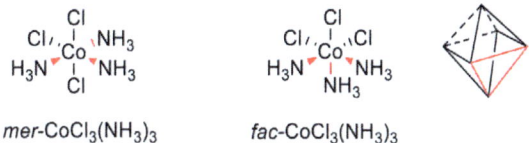

cis-[CoCl$_2$(NH$_3$)$_4$]$^+$ *cis*-platin   *cis*-but-2-ene
                                                       (Z-but-2-ene)

trans-[CoCl$_2$(NH$_3$)$_4$]$^+$ *trans*-platin   *trans*-but-2-ene
                                                       (E-but-2-ene)

Figure 7.9   *Cis* and *trans* octahedral MA_2B_4 complexes.

For a six-coordinate complex with two generic ligands, A and B, there is a final possible molecular formula, MA_3B_3. If you have made a model of *trans*-[CoCl$_2$(NH$_3$)$_4$]$^+$ just swap one of the remaining ammine ligands for another chloride. Though you have four ammine ligands to choose from, you will end up with the same product whichever you replace. The three Cl ligands lie in one plane (that includes Co) and the three B ligands in another, perpendicular plane (that also includes Co). This is the **mer** isomer. It's short for **meridional** (Figure 7.10).

mer-CoCl$_3$(NH$_3$)$_3$ *fac*-CoCl$_3$(NH$_3$)$_3$

Figure 7.10   Octahedral MA_3B_3 complexes can have either *mer* or *fac* geometry.

But if we take our *cis*-[CoCl$_2$(NH$_3$)$_4$]$^+$ model and do the same thing, we have two possible outcomes. We either get the same *mer* isomer as before, or we can create a new complex where all the A ligands are *cis* to each other and all the B ligands are *cis* to each other. This *fac* isomer has the A ligands on one face of the octahedron and the B ligands on the opposite face, hence the descriptor *fac* for **facial** (Figure 7.10).

Now let's just check you are visualising the *mer* and *fac* isomers correctly. So, going back to Figure 7.10, do either or both of the *mer* and *fac* isomers have planes of symmetry?

For the *mer* isomer, the thing to do is focus on pairs of identical ligands that are *trans* to each other. If you then draw a plane including the cobalt and the other four ligands, it will be a mirror plane, since it has (a) an NH$_3$ front left and behind right, or (b) a chloride above and below (Figure 7.11).

> If you are struggling to see why the *fac* isomer is called 'facial', copy the black, red and dashed octahedron and add NH$_3$ ligands to the red vertices and Cl$^-$ ligands to the vertices with dashed lines. You'll see how each set of three identical ligands is grouped round a *face* of the octahedron.

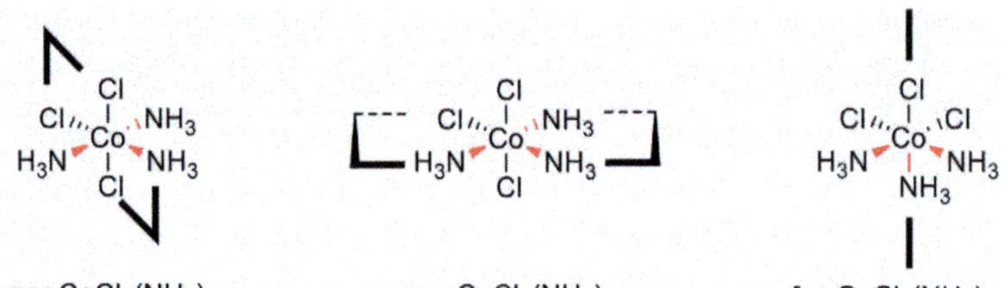

$mer\text{-}CoCl_3(NH_3)_3$ $mer\text{-}CoCl_3(NH_3)_3$ $fac\text{-}CoCl_3(NH_3)_3$

Figure 7.11 Checkpoint solution.

> ⚑ **Checkpoint**
>
> You should now be able to recognise and name *cis* and *trans* and *mer* and *fac* isomers of octahedral complexes.

For the *fac* isomer, the mirror planes need to bisect both the triangle of three chlorides and the triangle of three ammine ligands. We've shown one of these here. There are two more.

Reassuringly, the *fac* and *mer* descriptors are the only completely new stereochemical idea we learn in this chapter. These isomers don't arise with carbon atoms because you need six ligands to make them. And while the next section on chiral octahedral complexes may seem tricky, in fact we're just applying concepts we have come across before with organic chemicals.

7.5 Stereochemistry of Chiral Six-coordinate Complexes

To start this section, let's just check you understand the terms **monodentate**, **bidentate** and **tridentate** that we use to describe ligands in metal complexes. A monodentate ligand occupies just one coordination site. Ammine and chloro are monodentate ligands and we need six of them in total to form an octahedral complex. A bidentate ligand occupies two coordination sites and a tridentate ligand three sites.

For octahedral complexes that have three or more ligand types, chiral isomers are possible. Why not have a go with your models? An example with $MA_2B_2C_2$ is given in Figure 7.12. If you are feeling like a real challenge, you could look for all the 15 pairs of enantiomers that can arise with an MABCDEF complex!

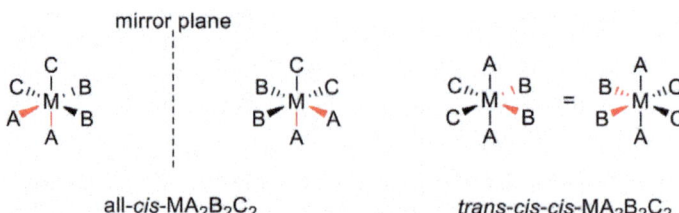

all-*cis*-$MA_2B_2C_2$ trans-*cis*-*cis*-$MA_2B_2C_2$

Figure 7.12 $MA_2B_2C_2$ complexes. The all-*cis* isomer is chiral.

The situation becomes much simpler with bidentate ligands. There is just one pair of enantiomers (Figure 7.13).

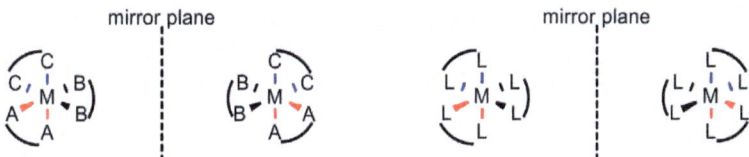

Figure 7.13 Octahedral complexes with three different, or three identical bidentate ligands.

At first sight, you may not see that these are enantiomers. In fact, we have already met this phenomenon in Chapter 6, with help from a snail and its helical shell (Figure 7.14). And, of course, as we learnt then, helices are chiral.

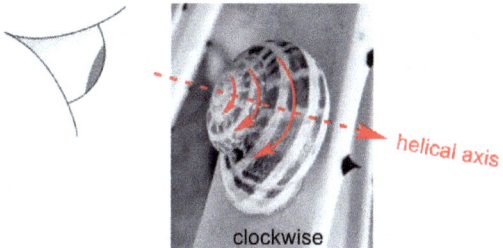

Figure 7.14 Helical snail shell.

These octahedral complexes with bidentate ligands have a stereogenic axis, like a screw thread and like a chiral helicene, biphenyl or allene. So all we need to do to describe the chirality is see which way the screw thread turns. Happily, the representation we have used here makes this rather simple. We focus on the ligands that are nearest to us in the 2D representation (marked *), the ones on the bold wedge bonds, and describe whether the bidentate links from the starred Ls to their linked Ls turn clockwise or anticlockwise (Figure 7.15).

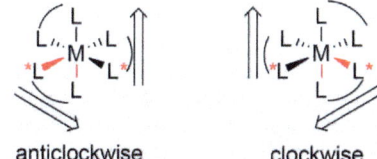

anticlockwise   clockwise

Figure 7.15 Assigning the configuration of chiral octahedral complexes.

Now we just need to note that for these octahedral complexes, the clockwise helix is called the delta Δ enantiomer and the anticlockwise helix is the lambda Λ enantiomer.

To check you've got this, let's assign the Δ or Λ configuration to this enantiomer of tris(acetylacetonato)iron(III), Fe(acac)$_3$ (Figure 7.16). Also, try to draw the enantiomer.

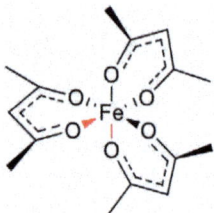

Figure 7.16   An enantiomer of Fe(acac)$_3$.

Again, focus on the oxygen atoms coming towards you at the ends of the wedged bonds and follow the links of the acac ligands (Figure 7.17).

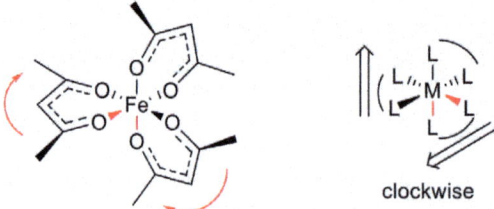

clockwise

Figure 7.17   (Δ)-Fe(acac)$_3$ showing the clockwise orientation of ligands.

That's right. It's the Δ enantiomer. To find the Λ enantiomer, just carefully draw the mirror image. Finally, we should note that the bidentate ligands we have considered so far are symmetrical A–A type. If we have A–B bidentate ligands, we still assign Δ and Λ in just the same way. However, we also need to decide whether our stereoisomer is *fac* or *mer*. Please have a go for the complex in Figure 7.18.

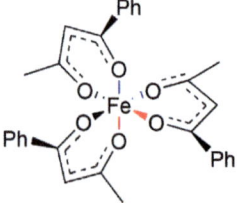

Figure 7.18   Is this the *fac* or *mer* isomer?

That's right. It's the enantiomer of the *fac* isomer, with all the Ph groups pointing towards us from the same face and the ligands describing anticlockwise arcs away from them.

7.6 Conclusion

Well done! At the end of this chapter, you should understand stereochemistry in square planar complexes and be able to name *cis* and *trans* isomers. You should be confident about recognising and naming *cis* and *trans* and *mer* and *fac* isomers of octahedral complexes and assigning Δ or Λ configurations to chiral octahedral complexes with bidentate ligands.

> ■ **Checkpoint**
>
> You should now be able to assign Δ or Λ configurations to chiral octahedral complexes with bidentate ligands.

This is also the end of the book. We warned you that stereochemistry is a challenging topic! It really is and you will need to keep on practising. We also promised that it would be a rewarding one. Our hope is that you are better able to visualise the 3D chemical world around you. Your new understanding will help you explore how 3D molecules come together to build complex structures, or react with each other to produce important new products. With this book and your model kit to hand, you should be well-equipped.

Exercises

1. Cisplatin is readily converted to another widely used chemotherapy drug, carboplatin.

People do not usually prefix the name carboplatin with *cis* or *trans*, but which is it? And why do you think the other stereoisomer cannot be prepared?

2 An isomer of oxaliplatin, an important chemotherapy drug, is shown. Identify and assign all the stereochemical features of this molecule.

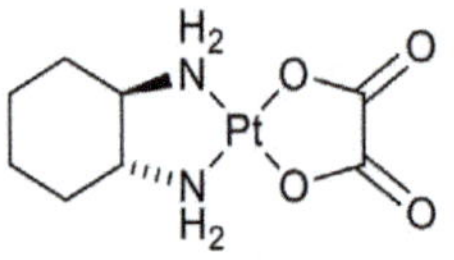

oxaliplatin

3. Is the isomer of RhPy$_3$Cl$_3$ shown the *fac* or *mer* isomer? Now draw the other isomer.

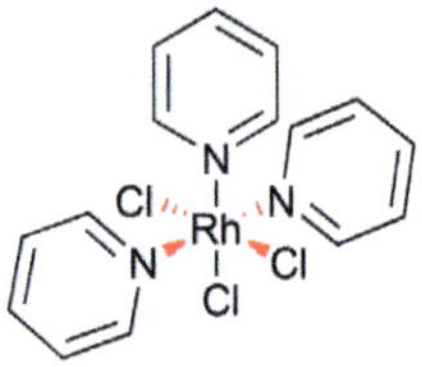

4. Tris(bipyridine)ruthenium(II) chloride has been proposed for applications in optical chemical sensors. Assign the absolute configuration of the enantiomer of the dication shown and draw the other isomer.

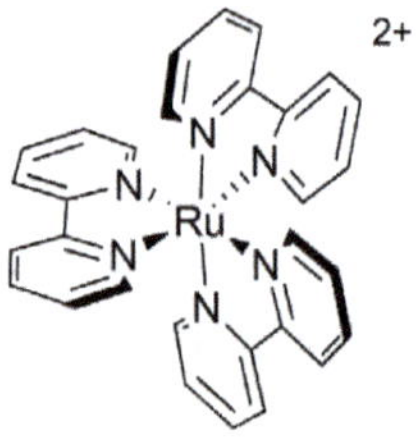

Answers

1. Whether you look at the ammine (NH$_3$) ligands or the bidentate dicarboxylate, you can see that the identical ligands are *cis* to each other. The bidentate ligand forms a six-membered ring when it binds to platinum. The bond lengths and angles mean it is impossible for this ligand to occupy two *trans* sites.
2. Whether you look at the bidentate diamine or the bidentate oxalate ligand, these again occupy *cis* positions, so this is a *cis* complex. The diamine has two stereogenic centres, both with *R* configuration.

3. The rhodium and the three nitrogen atoms lie in one plane. The rhodium and the three chlorides lie in another. This is the *mer* isomer. The *fac* isomer doesn't have any of the same ligands *trans* to each other, so you need to swap one of the two *trans*-pyridines with one of the two *trans*-chlorides so that each group of three is on a face of the octahedron.

e.g. swap these two

fac isomer

4. As ever, make sure that your complex is drawn in this orientation (it's hard to draw them any other way!) and focus on the two ligating atoms that are coming out towards you. Then draw arrows towards the other ends of the bidentate ligands. In this, case they go clockwise, so it's the Δ enantiomer. To find the Λ enantiomer, just draw the mirror image.

Index